WRASSE:
BIOLOGY AND USE
IN AQUACULTURE

WRASSE:
BIOLOGY AND USE
IN AQUACULTURE

Edited by

M.D.J. SAYER
Dunstaffnage Marine Laboratory,
Oban, UK

J.W. TREASURER
Marine Harvest McConnell,
Lochailort, UK

M.J. COSTELLO
Environmental Sciences Unit,
Trinity College, Dublin, Ireland

Fishing News Books

Copyright © 1996

Fishing News Books
A division of Blackwell Science Ltd
Editorial Offices:
Osney Mead, Oxford OX2 0EL
25 John Street, London WC1N 2BL
23 Ainslie Place, Edinburgh EH3 6AJ
238 Main Street, Cambridge,
  Massachusetts 02142, USA
54 University Street, Carlton,
  Victoria 3053, Australia

Other Editorial Offices:
Arnette Blackwell SA
  1, rue de Lille, 75007 Paris
  France

Blackwell Wissenschafts-Verlag GmbH
  Kurfürstendamm 57
  10707 Berlin, Germany

  Zehetnergasse 6, A-1140 Wien
  Austria

First published 1996

Set in 10 on 13 pt Times
by Best-set Typesetter Ltd., Hong Kong
Printed and bound in Great Britain by The
University Press, Cambridge

DISTRIBUTORS

Marston Book Services Ltd
PO Box 87
Oxford OX2 0DT
(*Orders:* Tel: 01865 206206
         Fax: 01865 721205
         Telex: 83355 MEDBOK G)
USA
Blackwell Science, Inc.
238 Main Street
Cambridge, MA 02142
(*Orders:* Tel: 800 215-1000
         617 876-7000
         Fax: 617 492-5263)

Canada
Oxford University Press
70 Wynford Drive
Don Mills
Ontario M3C 1J9
(*Orders:* Tel: 416 441 2941)

Australia
Blackwell Science Pty Ltd
54 University Street
Carlton, Victoria 3053
(*Orders:* Tel: 03 9347-0300
         Fax: 03 9349 3016)

A catalogue record for this book is
available from the British Library
ISBN 0-85238-236-7

Contents

Preface

Sea lice (Copepoda, Caligidae) are ectoparasitic copepods which infest salmonids of the northern hemisphere. The lice feed on the mucus, skin and blood of the host and, if they are not removed, cause open wounds exposing the fish to osmotic stress and secondary infections. In sea cage rearing of Atlantic salmon (*Salmo salar* L.) repeated lice infestations are a major problem to the industry. For many years, salmon farmers have controlled sea lice by bathing the cages with organophosphate pesticides. Regular treatments are necessary once sea lice populations become established because the larval stages are not affected, and the sea lice reproduce rapidly. As well as being expensive (due to continual retreatment) and laborious to use, pesticide use can cause stress to the salmon and reduce growth rates. There is also potential risk of killing fish during treatment, and reduced sensitivity of lice through the use of organophosphates over many years has led to partial failure of treatments.

Environmental and financial concerns over widespread chemical use in the salmon farming industry resulted in a number of biological and chemical alternatives being investigated. One successful alternative sea-lice treatment involves cleaning species where one species of fish (the cleaner) feeds on parasites from another species (the host). As potential cleaner species, the north European wrasses (family Labridae) attracted most attention, because natural cleaning behaviour had been observed for some of the species. Experimental trials in the late 1980s by Åsmund Bjordal, a fish biologist working in Norway, confirmed that certain wrasse species were effective in the control of sea lice infestation in commercial salmon farming. Following publication of these initial findings, the use of wrasse as cleaners spread rapidly throughout the salmon culture industry, and is now a widespread primary method of sea-lice control on many farms in Scotland, Norway, and, to a lesser extent, Ireland.

The application of wrasse as cleaners also stimulated scientific interest in a fish group largely ignored prior to their use in salmon farms. Much of this research was fundamental in nature, but in many cases also industry-led. Salmon growers' associations commissioned biological, ecological, behavioural, physiological and parasitological studies in order to investigate natural population distributions and dynamics of the wrasse; to improve cleaning efficiency and wrasse survival in salmon cages; to assess potential disease threats from the introduction of wild-caught fish into the culture situation; and to examine wrasse culture in general. The wrasse fishery is unusual in north temperate waters in that it is trap-orientated, therefore documentation of the fishery and assessment of the consequences on the stock were required.

Although the technique of using wrasse as cleaners had spread rapidly through the salmon farming industry, interest in the application was beginning to diminish after four or five years in Scotland and Ireland. To some extent, much of this decline resulted from perceived limitations of the technique, and to a lack of substantiated information relating to the practical requirements of wrasse use. Anecdotal evidence and rumour began to work against this biological methodology, always considered by some as slightly eccentric in an industry conditioned to vaccines and pesticides.

Some reports on wrasse were published in the scientific and trade literature, but what appeared to be lacking was a forum whereby researchers from academic institutions and interested parties from the salmon farming industry could exchange views and experiences on past usage, and discuss the future of the technique and its possible expansion to other cultured species. It was for this reason that an international symposium under the title 'Wrasse Biology and Aquaculture Applications' was held in October 1994, at Dunstaffnage Marine Laboratory near Oban on the west coast of Scotland. The symposium attracted over 60 delegates, mainly from Scotland, Norway and Ireland, but also from England, Portugal and Chile. This book constitutes the main proceedings of that symposium and includes most of the presentations given over the two days.

It was always the intention that this book should provide a useful reference for both the scientific and fish farming communities. All of the included papers have been refereed in order to provide a consistent quality of presentation. Because of funding limitations, or the practicalities of executing trials in a commercial situation, results of some studies are provisional rather than scientifically conclusive. However, these studies offer practical knowledge of considerable worth to the salmon farming industry, and one of the purposes of a book dealing exclusively with wrasse was to bring into the wider domain information which may not have been published elsewhere.

Both the book and the symposium received substantial support from Highland and Islands Enterprise, the Crown Estate, the Fisheries Society of the British Isles, Marine Harvest, the Scottish Association for Marine Science, Dunstaffnage Marine Laboratory, Argyll and Islands Enterprise, the Scottish Salmon Growers Association and BOCM Pauls. Without such support neither venture would have proceeded.

Martin Sayer, Dunstaffnage Marine Laboratory, Oban
Jim Treasurer, Marine Harvest McConnell Ltd, Lochailort
Mark Costello, Environmental Sciences Unit, Trinity College, Dublin

December 1995

PART I
WRASSE BIOLOGY

Chapter 1
North European wrasse: identification, distribution and habitat

M.D.J. SAYER[1] and J.W. TREASURER[2] *¹ Centre for Coastal and Marine Sciences, Dunstaffnage Marine Laboratory, P.O. Box 3, Oban, Argyll PA34 4AD, UK; and ² Marine Harvest McConnell, Lochailort, Inverness-shire PH38 4LZ, UK*

A brief account is given of the five species of wrasse most commonly found in north European waters: goldsinny (*Ctenolabrus rupestris*), rock cook (*Centrolabrus exoletus*), corkwing wrasse (*Crenilabrus melops*), ballan wrasse (*Labrus bergylta*) and cuckoo wrasse (*Labrus mixtus*). Notes on identification, distribution and habitat are included for each species. The less common wrasse species, occasionally recorded in coastal waters of the north-east Atlantic (scale-rayed wrasse, *Acantholabrus palloni*; Baillon's wrasse, *Crenilabrus bailloni*; rainbow wrasse, *Coris julis*), are mentioned in brief.

1.1   Introduction

The wrasse family (division Teleostei, order Perciformes, family Labridae) is the second largest of marine fishes and the third largest perciform family, with at least 60 genera (Nelson, 1994). Members are widely distributed, being common through temperate and tropical seas (Wheeler, 1969; Wernerus, 1989). Eight wrasse species have been reported for north European waters, five of which are common inshore fishes (Lythgoe & Lythgoe, 1991): goldsinny [*Ctenolabrus rupestris* (L.)], rock cook [*Centrolabrus exoletus* (L.)], corkwing wrasse [*Crenilabrus melops* (L.)], ballan wrasse (*Labrus bergylta* Ascanius), and cuckoo wrasse [*Labrus mixtus* (L.)]. There are rare and infrequent records of the other three species: scale-rayed wrasse [*Acantholabrus palloni* (Risso)], Baillon's wrasse [*Crenilabrus bailloni* (Valenciennes)], and rainbow wrasse [*Coris julis* (L.)] (Wheeler, 1992). This brief review gives guidance on the identification of different wrasse species, describes the likely distribution and preferred habitat of each species. More detailed reviews of the biology of northern European wrasse have been given by Costello (1991) and Darwall *et al.* (1992).

1.2   Goldsinny

Goldsinny are sometimes referred to as Jago's goldsinny (Wheeler, 1969; Sjölander, 1972). The established scientific name for goldsinny is *Ctenolabrus rupestris* (Linnaeus, 1758; Wheeler, 1992); a synonym sometimes used is *Cteno-labrus suillus* (L.) (Lythgoe & Lythgoe, 1991). Common European names for the goldsinny are: cténolabre or rouquié (France), grivieta (Spain), klip-lipvisch (Holland), klippenbarsch (Germany), havkaruds (Denmark), bergnaeb or raade (Norway) and bergsnultra (Sweden) (Wheeler, 1969; Quignard & Pras, 1986). Goldsinny are relatively small fish, usually 100–120 mm total length, with maxima in the range of 159–180 mm (Smitt, 1892; Quignard & Pras, 1986; Sayer *et al.*, 1995). Maximum ages for males and females are 14 and 20 years respectively (Sayer *et al.*, 1995).

The adults are orange-red in colour; juveniles may be dull green (Wheeler, 1969). Male goldsinny may have red spots on their flanks at all times of the year (Sjölander *et al.*, 1972; Hilldén, 1981), or gold-coloured flanks during the repro-ductive period (Costello, 1991). Females can be distinguished by vertical bands in the mid-ventral region and a black cloaca (Hilldén, 1981). However, practical experience of attempting to visually sex goldsinny proved unreliable (Sayer *et al.*, 1993). The most distinctive feature of goldsinny appearance is a black patch on the dorsal edge of the caudal peduncle close to the caudal fin (Fig. 1.1). A second dark patch is sometimes present on the dorsal fin extending for the first four or five spines (Wheeler, 1969; Fig. 1.1).

Postlarvae settling in the shallow rocky subtidal (late August in Scotland–Sayer *et al.*, 1993) do not possess the tail fin spot, which appears *c.* three to four days post-settlement (Smitt, 1892; Sayer, *pers. obs.*). During the first year, four colour morphs have been observed for juvenile goldsinny. Most common is possession of a single horizontal white or golden stripe extending from above the mouth, around the eyes, to immediately posterior of the opercular opening (Chapter 12, this volume). This horizontal stripe is sometimes present along with vertical stripes on the body, although the vertical stripes are often present on their own (Sayer, *pers. obs.*). Some juvenile goldsinny possess neither horizontal or vertical stripes.

Goldsinny are recorded throughout the Mediterranean and in the eastern Black Sea (Quignard & Pras, 1986). They are common along inshore eastern Atlantic coasts from Morocco to Norway, being found also in the English Channel, the

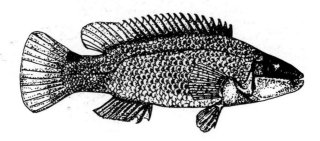

Fig. 1.1 Goldsinny, *Ctenolabrus rupestris* (redrawn from Smitt, 1892).

North Sea and the Baltic (Wheeler, 1969; Quignard & Pras, 1986; and Chapter 6, this volume). The typical habitat for goldsinny is on rocky shores, often with the requirement of shelter holes (Costello, 1991; Sayer *et al.*, 1993); they are rarely caught over sand (Wheeler, 1969; Sayer, *unpubl. data*). Goldsinny are found in the intertidal (in summer) down to depths of 50 m (Quignard & Pras, 1986; Sayer *et al.*, 1993), with adults found deeper than young (Quignard & Pras, 1986; Sayer *et al.*, 1993).

1.3   Rock cook

Rock cook are sometimes called small-mouthed wrasse (Wheeler, 1969, 1992). The established scientific name is *Centrolabrus exoletus* (Linnaeus, 1758) (Wheeler, 1992); there are no common synonyms (Quignard & Pras, 1986). European names for rock cook are: petite vieille (France), graesraade (Norway), Småmundet gylte (Denmark) and Smämunte snultra (Sweden) (Wheeler, 1969; Quignard & Pras, 1986). Rock cook are similar in size to goldsinny, most commonly in the length range 100–140 mm (Smitt, 1892; Chapter 7, this volume) and a maximum length of 165 mm (Chapter 2, this volume). Maximum age for rock cook in Scotland has been reported as nine years (Treasurer, 1994a).

Rock cook are usually red-brown on the back, but lighter on the belly (Wheeler, 1969; Quignard and Pras, 1986). The body sides and lower part of the head are yellow-orange (Quignard & Pras, 1986). The lower part of the head has blue and red stripes. Scales, particularly on the back, are partly iridescent blue-purple (Lythgoe & Lythgoe, 1991). The most obvious distinguishing feature of the rock cook are two dark, broad bands, one at the base of the caudal fin, and one on the fin (Fig. 1.2). There is no reliable method for distinguishing male and female rock cook visually: although there is a possible tendency for males to possess more iridescent blue colouration, this is not always dependable.

There is little detailed information relating to rock cook distribution. They do not occur in the Mediterranean, and are only found in large numbers north of the

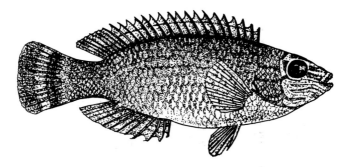

Fig. 1.2   Rock cook, *Centrolabrus exoletus* (redrawn from Smitt, 1892).

Bay of Biscay (Lythgoe & Lythgoe, 1991). They are found on the west coast of France and on most west coasts of the British Isles, but may be absent in the eastern English Channel and southern North Sea (Wheeler, 1969; Lythgoe & Lythgoe, 1991). Rock cook are found on the southern coasts of Norway and Sweden, but are absent from much of the Baltic (Wheeler, 1969). There are reports of rock cook in Greenland (Smitt, 1892; Quignard & Pras, 1986).

Rock cook are usually associated with rocky coasts, although occasional specimens have been captured in seine nets on sandy beaches (Sayer, *pers. obs.*). They have been observed at depths below 40 m (Sayer, *pers. obs.*) and have also been caught in intertidal rock pools (Wheeler, 1969). There may be a tendency for rock cook to move into deeper water during the winter, although some individuals have been captured in shallow water at that time of year (Sayer *et al.*, 1994).

1.4   Corkwing wrasse

All recent references to this species call it corkwing wrasse, although Smitt (1892) referred to it as the gilt-head or connor. The established scientific name is *Crenilabrus melops* (Linnaeus, 1758; Wheeler, 1992) although the synonym *Symphodus melops* (L.) is frequently used (Lythgoe & Lythgoe, 1991). The common names for corkwing used in other European countries are: tordo (Spain), crénilabre (France), zwartoog-lipvisch (Holland), schwarzäugiger lippfisch (Germany), savgylte (Denmark), grönaade (Norway) and skårsnultra (Sweden) (Wheeler, 1969; Quignard and Pras, 1986). Maximum corkwing total lengths of between 235 mm and 280 mm have been reported (Wheeler, 1969; Quignard & Pras, 1986; Darwall *et al.*, 1992; Treasurer, 1994a). A detailed examination of 278 corkwing caught on the west coast of Scotland recorded a maximum length and age of 212 mm and six years (Chapter 2, this volume). Darwall *et al.* (1992) give a maximum corkwing age of nine years.

The most distinctive feature of the corkwing is a dark spot on the caudal peduncle either on, or slightly below the lateral line (Lythgoe & Lythgoe, 1991; Fig. 1.3). Both sexes have a dark crescent-shaped patch behind the eye (Wheeler, 1969; Costello, 1991), though this is sometimes not very evident (Quignard & Pras, 1986). Colour can vary significantly with sex, season and maturity. However, in general the body colouring of females and juveniles tends to be dull green-brown, whereas males are often more red-brown with distinctive blue and red stripes on the lower head and gill covers (Quignard & Pras, 1986; Lythgoe & Lythgoe, 1991). Mature females have a conspicuous dark blue urogenital papilla (Quignard & Pras, 1986; Costello, 1991). Visual sexual differentiation can be misleading due to the presence of satellite or type-2 males. These are functional males but with a female appearance and abdominal swelling during the spawning season (Dipper & Pullin, 1979). They may make up at least 10% of the male population (Chapter 12, this volume).

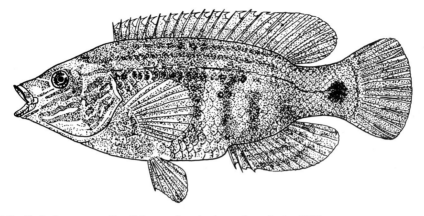

Fig. 1.3 Corkwing wrasse, *Crenilabrus melops* (redrawn from Smitt, 1892).

Corkwing are found in the eastern Atlantic from Morocco to half-way up the Norwegian coast, on the Azores, and well into the Baltic (Quignard and Pras, 1986). Their distribution within the Mediterranean appears limited to the north-west quarter (Quignard & Pras, 1986). Although usually associated with rocky shores, corkwing are typically found in areas of high algal cover, e.g. kelp forest, eel-grass beds (Quignard and Pras, 1986; Lythgoe & Lythgoe, 1991; Sayer, *pers. obs.*). They are shallow-water fishes commonly found in intertidal rockpools (Wheeler, 1969; Lythgoe & Lythgoe, 1991) and in water depths of less than 5 m (Costello, 1991), although they can occur in depths of 15–18 m (Quignard, 1966; Costello, 1991; Sayer *et al.*, 1994).

1.5 Ballan wrasse

This wrasse species is always referred to as ballan wrasse in recent literature. The established name is *Labrus bergylta* Ascanius, 1767 (Wheeler, 1992). Wheeler (1969) gives *Labrus maculatus* as a synonym. Common European names for the ballan are: vielle commune or labre (France), vaqueta (Spain), gevlekte lipvisch (Holland), gefleckter lippfisch (Germany), and berggylt (Denmark, Norway, Sweden) (Smitt, 1892; Wheeler, 1969; Quignard & Pras, 1986). Ballan are the largest of the north European wrasses, and may attain a total length of 600 mm (Quignard & Pras, 1986; Darwall *et al.*, 1992), though lengths of 300–500 mm are more common (Quignard & Pras, 1986; Lythgoe & Lythgoe, 1991). A range of 125–280 mm total length was reported for a sample of 131 ballan caught on the west coast of Scotland (Treasurer, 1994b). A maximum age of 25 years has been reported for ballan (Darwall *et al.*, 1992).

Body colouring can be so variable that it is not possible to give a general description for the species. Ballan colouration is probably not dependent on season or sex (Smitt, 1892), and may possibly change to match the background

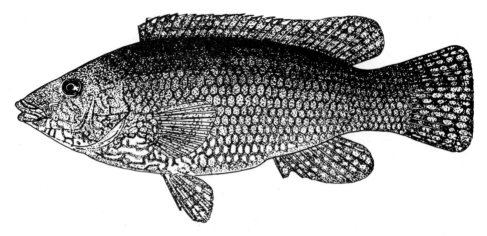

Fig. 1.4   Ballan wrasse, *Labrus bergylta* (redrawn from Smitt, 1892).

colour of its immediate surroundings (Wheeler, 1969). The body scales are large, often with dark borders, which makes them appear very prominent (Fig. 1.4). Red sinuous markings are typical on the lower part of the head, opercular covers, and area below the pectoral fins (Fig. 1.4). Juvenile ballan (80–120 mm) may be a light green (Costello, 1991; Lythgoe & Lythgoe, 1991; Treasurer, *pers. obs.*).

Ballan are recorded on eastern Atlantic coasts from Morocco to Norway, including Madeira, the Azores and the Canaries (Quignard and Pras, 1986). They are found in the North Sea and western parts of the Baltic (Wheeler, 1969), but may be rare or even absent from the Mediterranean (Quignard & Pras, 1986; Lythgoe & Lythgoe, 1991). Ballan occur in rocky areas typically inhabited by goldsinny and rock cook. Juveniles are sometimes found in the intertidal (Quignard & Pras, 1986); adult presence may extend below 30 m (Sayer, *pers. obs.*).

1.6   Cuckoo wrasse

This wrasse species is always referred to as cuckoo wrasse in recent literature (Wheeler, 1992); it was called the striped wrasse by Smitt (1892). Wheeler (1992) gives a detailed explanation for establishing the scientific name for cuckoo as *Labrus mixtus* Linnaeus, 1758. *Labrus bimaculatus* is a common synonym, *Labrus ossifagus* less so. Common European names for cuckoo are vielle coquette or labre mêlé (France), gallano (Spain), lippfisch (Germany), blåstak (male) and rödnaeb (female) (Denmark and Norway), blåstål (Sweden) (Wheeler, 1969; Quignard and Pras, 1986). The maximum length and age for cuckoo is 350 mm and 17 years respectively (Darwall *et al.*, 1992).

In the mature fish the sexes differ markedly in colour and pattern. The body of females and immature males is red-brown to orange in colour with very conspicuous dark and light blotches along the back and dorsal edge of the caudal peduncle (Lythgoe & Lythgoe, 1991; see also Fig. 1.5). Much of the male body, including dorsal and caudal fins, is a brilliant orange or orange-yellow colour with bright blue stripes and patches (Lythgoe & Lythgoe, 1991; see also Fig. 1.6). The anterior portion of the body and the head are a mixture of iridescent blue and gold markings. When maturing into a male, the colour changes from the head backwards. It is not uncommon to see cuckoo with the blue head colouration of the male, and the female orange body with black and white markings. The sexual status of the fish with such colouration is not known at present.

Cuckoo wrasse are distributed along the coasts of the eastern Atlantic from Senegal in the south to Norway in the north, including Madeira and the Azores (Quignard & Pras, 1986). It is found throughout the Mediterranean but is largely absent from the Baltic (Wheeler, 1969; Lythgoe & Lythgoe, 1991). Cuckoo inhabit a large depth range of perhaps 2–200 m, (Quignard & Pras, 1986), though are seldom found in water depths of less than 10 m (Wheeler, 1969; Lythgoe &

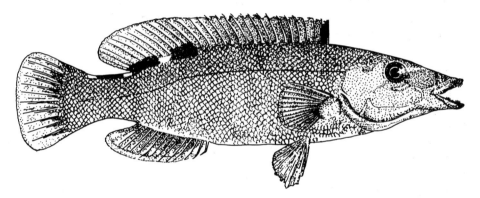

Fig. 1.5   Female cuckoo wrasse, *Labrus mixtus* (redrawn from Smitt, 1892).

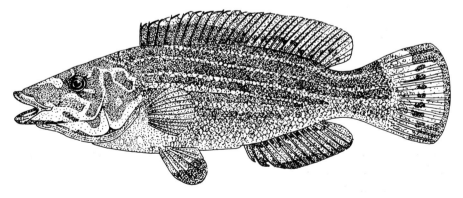

Fig. 1.6   Male cuckoo wrasse, *Labrus mixtus* (redrawn from Smitt, 1892).

Lythgoe, 1991). Again, they are associated mainly with rocky shores (Wheeler, 1969).

1.7   Scale-rayed wrasse

The established scientific name for the scale-rayed wrasse is *Acantholabrus palloni* (Risso, 1810) (Wheeler, 1992). It has no common synonyms, but other common European names for it are: roucaou (France), tac rocas (Spain) and brunsnultran (Sweden) (Smitt, 1892; Quignard & Pras, 1986). It is green-brown on the back with indistinct clear bands which may be red on the sides, but light on the belly (Wheeler, 1969). There are two dark blotches, one on the upper edge of the caudal peduncle, the other on the dorsal fin at the junction of the spiny and soft rays (Wheeler, 1969; Quignard & Pras, 1986). Although it has a distribution from Norway to Cape Lopez, around Madeira and the Azores, and in the Mediterranean, it is nowhere common (Wheeler, 1969; Quignard & Pras, 1986). Consequently, little is known of its biology. It is believed to be a solitary fish living on rocky and sandy bottoms in depths of 60–250 m in the Mediterranean, and 18–55 m off the Norwegian coast (Wheeler, 1969; Quignard and Pras, 1986). Adult specimens of scale-rayed wrasse have measured 200–280 mm in length, but there are records of this species at less than 50 mm long (Wheeler, 1969).

1.8   Baillon's wrasse

The established scientific name for Baillon's wrasse is *Crenilabrus bailloni* Valenciennes, 1839 (Wheeler, 1992) although the synonym *Symphodus bailloni* is sometimes used (Quignard & Pras, 1986). European names for it are vielle (France) and tort (Spain) (Quignard & Pras, 1986). Both sexes have a dark spot on the caudal peduncle and another dark blue spot on the beginning of the soft rays of the dorsal fin (Quignard & Pras, 1986). There are often five vertical dark brown patches on the upper part of the flank, sometimes reaching the belly and anal fin (Lythgoe & Lythgoe, 1991). The juveniles and females are brown with the adult females possessing a dark grey urogenital papilla (Quignard & Pras, 1986). The males can be green, red or bright brown, with a brown-green head with orange stripes (Quignard & Pras, 1986; Lythgoe & Lythgoe, 1991). Baillon's wrasse are usually in the 150–180 mm size range, with maxima approaching 200 mm (Quignard & Pras, 1986).

Baillon's wrasse are distributed in the eastern Atlantic from the North Sea to Mauritania, and off the coast of Spain and the Balearic islands in the Mediterranean (Quignard & Pras, 1986; Lythgoe & Lythgoe, 1991). They have been captured on the coast of Jersey and on the Galway coast of Ireland (Wheeler & Clark, 1984). Several specimens have also been found in the eastern North Sea (Nijssen & de Groot, 1974). This species is found over rocks and seagrass beds from the surface down to 50 m (Quignard & Pras, 1986; Lythgoe & Lythgoe, 1991).

1.9   Rainbow wrasse

The established scientific name for the rainbow wrasse is *Coris julis* (Linnaeus, 1758) (Wheeler, 1992). There is no common synonym and other European names for it are girelle (France) and julia (Spain) (Quignard & Pras, 1986). Colour varies with sex, age and depth. Males have an orange-red and black spot on the first three dorsal fin rays (Quignard & Pras, 1986). They may be blue, olive-green, brown or grey on the dorsal surface with a white belly, and flanks with zig-zag orange or red stripes (Lythgoe & Lythgoe, 1991). Females and juveniles in shallow water have a brownish back (red in deeper water), and white-yellow belly, with a large white stripe on the flanks. A black spot is present on the posterior edge of the operculum in both sexes and juveniles.

Although the rainbow wrasse is rare in north Europe, it is widely distributed in all parts of the Mediterranean, infrequently in the Black Sea, and is also found off Spain, Portugal, the Azores, Madeira, and the Canaries (Quignard & Pras, 1986). There are four nineteenth-century records of this species on the south coast of Britain (Wheeler, 1992) and two instances of capture on the Scandinavian coast (Smitt, 1892). Rainbow wrasse may be solitary or found in small shoals. They are very abundant around sea-grass and weed-covered rocks in the Mediterranean (Lythgoe & Lythgoe, 1991), and are reported down to depths of 120 m, with males in deeper water (Quignard & Pras, 1986). Rainbow wrasse bury themselves in sand at night and also if alarmed (Quignard & Pras, 1986; Videler, 1988). Males reach up to 250 mm total length compared with 180 mm in females (Quignard & Pras, 1986).

References

Costello, M.J. (1991) Review of the biology of wrasse (Labridae: Pisces) in northern Europe. *Progress in Underwater Science* **16**: 29–51.

Darwall, W.R.T., Costello, M.J., Donnelly, R. and Lysaght, S. (1992) Implications of life-history strategies for a new wrasse fishery. *Journal of Fish Biology* **41** (Supplement B): 111–23.

Dipper, F.A. and Pullin, R.S.V. (1979) Gonochorism and sex-inversion in British Labridae (Pisces). *Journal of Zoology* (London) **187**: 97–112.

Hilldén, N.-O. (1981) Territoriality and reproductive behaviour in the goldsinny, *Ctenolabrus rupestris* L. *Behavioural Processes* **6**: 207–21.

Lythgoe, J.N. and Lythgoe, G.I. (1991) *Fishes of the Sea: The North Atlantic and Mediterranean.* Blandford Press, London.

Nelson, J.S. (1994) *Fishes of the World* 3rd edn. John Wiley and Sons, Inc., New York.

Nijssen, H. and de Groot, S.J. (1974) Catalogue of the fish species of the Netherlands. *Beaufortia* **21**: 173–207.

Quignard, J.-P. (1966) Recherches sur les Labridae (Poissons, Teleostéens, Perciformes) des côtes européennes: systématique et biologie. *Naturalia Monspeliensia (Zoologie)* **5**: 7–248.

Quignard, J.-P. and Pras, A. (1986) Labridae. In *Fishes of the North-eastern Atlantic and the Mediterranean* Vol. II (Ed. by Whitehead, P.J.P., Bauchot, M.-L., Hureau, J.-C., Nielsen, J. and Tortonese, E.). Paris: UNESCO, pp. 919–42.

Sayer, M.D.J., Cameron, K.S. and Wilkinson, G. (1994) Fish species found in the rocky sublittoral during winter months as revealed by the underwater application of the anaesthetic quinaldine. *Journal of Fish Biology* **44**: 351–3.

Sayer, M.D.J., Gibson, R.N. and Atkinson, R.J.A. (1993) Distribution and density of populations of goldsinny wrasse (*Ctenolabrus rupestris*) on the west coast of Scotland. *Journal of Fish Biology* **43** (Supplement A): 157–67.

Sayer, M.D.J., Gibson, R.N. and Atkinson, R.J.A. (1995) Growth, diet and condition of goldsinny on the west coast of Scotland. *Journal of Fish Biology* **46**: 317–40.

Sjölander, S., Larson, H.O. and Engström, J. (1972) On the reproductive behaviour of two labrid fishes, the ballan wrasse (*Labrus bergylta*) and Jago's goldsinny (*Ctenolabrus rupestris*). *Revue Comporte Animale* **6**: 43–51.

Smitt, F.A. (1892) *A History of Scandinavian Fishes* 2nd edn. P.A. Norstedt and Söner, Stockholm.

Treasurer, J.W. (1994a) The distribution, age and growth of wrasse (Labridae) in inshore waters of west Scotland. *Journal of Fish Biology* **44**: 905–18.

Treasurer, J.W. (1994b) Distribution and species and length composition of wrasse (Labridae) in inshore waters of west Scotland. *Glasgow Naturalist* **22**: 409–17.

Videler, J.J. (1988) Sleep under sand cover of the labrid fish *Coris julis*. In *Sleep '86* (Koella, W.P., Obál, F., Schulz, H. and Visser, P., eds). Gustav Fischer Verlag, Stuttgart, pp. 145–7.

Wernerus, F. (1989) Etude des mécanismes sous-tendant les systèmes d'appariement de quatre espèces de poissons labridés méditerranéens des genres *Symphodus* Rafinesque, 1810 et *Thalassoma* Linné, 1758. *Cahiers d'Ethologie appliquée* **9**: 117–320.

Wheeler, A. (1969) *Fishes of the British Isles and North-East Atlantic*. Macmillan, London.

Wheeler, A. (1992) A list of the common and scientific names of fishes of the British Isles. *Journal of Fish Biology* **41** (Supplement A), 37 pp.

Wheeler, A. and Clark, P. (1984) New records for the occurrence of *Crenilabrus bailloni* (Osteichthyes: Perciformes: Labridae) in the waters of northern Europe. *Journal of the Marine Biological Association of the United Kingdom* **64**: 1–6.

Chapter 2
Seasonal, sexual and geographical variation in the biology of goldsinny, corkwing and rock cook on the west coast of Scotland

M.D.J. SAYER[1], R.N. GIBSON[1] and R.J.A. ATKINSON[2] [1] *Centre for Coastal and Marine Sciences, Dunstaffnage Marine Laboratory, P.O. Box 3, Oban, Argyll PA34 4AD, UK; and* [2] *University Marine Biological Station, Millport, Isle of Cumbrae KA28 0EG, UK*

A study of seasonal, geographical and sexual variations of growth, diet, somatic condition (K_S), gonadosomatic condition (GSI) and hepatosomatic condition (HSI), of goldsinny (*Ctenolabrus rupestris*), rock cook (*Centrolabrus exoletus*) and corkwing (*Crenilabrus melops*) wrasse was made on specimens caught on the west coast of Scotland. Capture throughout the year was by either baited creel, beam trawling or by anaesthetics applied underwater. Maximum age and total length for the three species were 20+ years and 159 mm for goldsinny ($n = 792$), 8+ years and 165 mm for rock cook ($n = 133$), and 6+ years and 212 mm for corkwing ($n = 278$).

Considerable variation in growth rates was recorded between sexes for goldsinny and corkwing, and between different locations for goldsinny. There were also differences in K_S, GSI, HSI and dietary trends for all three species associated with variation in season, sex and location. Comparisons with previously published growth data for all three species indicate considerable geographical variation. The present practice of removing wrasse from the fishery at and above 100 mm total length may have a differing impact on the sustainability of exploited populations depending on specific, sexual and geographical variation in growth rates.

2.1   Introduction

Goldsinny [*Ctenolabrus rupestris* (L.)], rock cook [*Centrolabrus exoletus* (L.)] and corkwing wrasse [*Crenilabrus melops* (L.)] are species of north European wrasses (Labridae) which are effective in the control of sea lice (Copepoda, Caligidae) infestations in sea cage farming of Atlantic salmon (*Salmo salar* L.) (for reviews see Bjordal, 1991; Costello, 1993; Treasurer, 1993). The financial and environmental benefits of using wrasse as cleaner-fish have resulted in substantial commercial exploitation of natural populations of these wrasse species in Norway,

Scotland, England and Ireland (Darwall *et al.*, 1992; Treasurer, 1994). It is not known whether these three species are capable of sustaining intensive fisheries. If the size of a sustainable fishery is to be estimated, then it is essential to obtain detailed biological information relating to the target species. Such information is lacking for many aspects of the biology of goldsinny, rock cook and corkwing wrasse (Costello, 1991; Darwall *et al.*, 1992). This account details the seasonal, geographical and sexual variations of growth, diet and condition (somatic, gonadal and hepatic) of goldsinny, rock cook and corkwing caught from selected areas on the west coast of Scotland.

2.2   Materials and methods

2.2.1   *Sampling methods and sites*

From May 1992 to June 1993, 792 goldsinny were obtained on a regular basis. Of this total, 475 were caught specifically for biological examination; the other 317 were obtained for other purposes but the sex, total length, standard length and total weight were recorded in most cases. From April to December, goldsinny were captured using traps consisting of modified prawn creels (see Treasurer, 1991, for methodology); they were baited with broken sea urchin, *Echinus esculentus* L. Outside this period, goldsinny rarely fed and were not attracted into creels. At these times they were therefore captured using anaesthetic applied underwater (Sayer *et al.*, 1994 give methodology). Collection using anaesthetic was also used to capture juvenile goldsinny that were neither attracted to, nor retained by, baited traps. In addition, a small number of goldsinny were purchased from a commercial supplier and occasional samples of goldsinny were captured in fyke nets. Goldsinny collection was concentrated at two areas on the west coast of Scotland: along the south-east coast of the Isle of Cumbrae (55°46'N, 4°53'W), and around Maiden Island (56°27'N, 5°25'W) near Oban (Fig. 2.1). Smaller numbers of goldsinny were obtained from the islands of Luing and Torsa (56°17'N, 5°34'W), Mull (56°30'N, 5°38'W) and Lismore (56°29'N, 5°28'W). Rock cook and corkwing were obtained on a regular basis from May 1992 to February 1994 by four methods. From April to December rock cook were captured using baited traps using the same methodology as used for goldsinny capture. Some rock cook were sampled using anaesthetic applied underwater (Sayer *et al.*, 1994) during the winter months. Almost all the samples of corkwing were captured using a 3 m beam trawl towed over weed-covered gravel/boulder fields in water depths of between 3 to 10 m. Occasionally corkwing were caught by creel. In addition, for length/age analyses, a number of rock cook were obtained from a commercial supplier who used fyke nets as a method of capture. Small numbers of rock cook were collected from the islands of Cumbrae and Luing and at sites in and close to Dunstaffnage Bay (56°28'N, 5°26'W). Nearly all corkwing specimens were obtained in the vicinity of Dunstaffnage Bay with occasional captures at

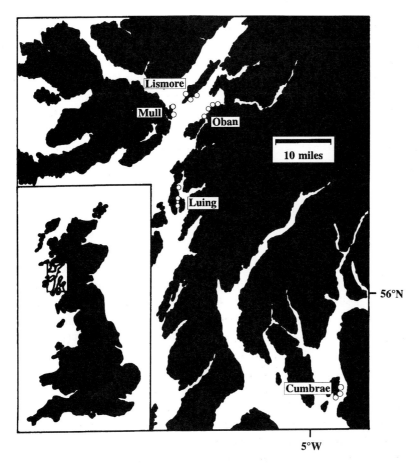

Fig. 2.1 Locations on the west coast of Scotland from where wrasse were sampled (redrawn from Sayer *et al.*, 1995). Inset of mainland UK illustrates map area.

Luing and Cumbrae. In total, 63 rock cook and 203 corkwing were collected for biological examination; other specimens from related experimental work were available for age/length determination.

2.2.2   *Age and growth studies*

Fish were killed by prolonged anaesthesia (benzocaine, ethyl p-amino benzoate; Sigma), and measured to the nearest 0.5 mm for standard length (SL) and total length (TL). After blotting dry on absorbent paper, the total weight (TW) of each specimen was obtained. The sagittal otoliths were removed and stored dry prior to being observed on a dark background using a stereomicroscope (Olympus SZ60) after first being cleaned in water, dehydrated in ethanol and cleared in creosote (Qasim, 1957). Growth was not back-calculated.

2.2.3   *Dietary analysis*

Goldsinny, rock cook and corkwing wrasse do not have differentiated stomachs (Quignard, 1966) and the gut contents were taken from the full length of the gut, transferred to water and identified under a stereomicroscope. Because some wrasse used for dietary determination were caught by baited trap, the time interval between entering the trap and eventual dissection could not be determined. Numerical, volumetric or gravimetric methods of gut contents analysis were therefore not appropriate. Instead, the presence of each group of prey organisms was recorded and the number of guts containing that group expressed as a percentage of the total number of guts containing food (Hyslop, 1980). This percentage occurrence method gives a relatively crude description of diet. However, providing that the major dietary constituents are relatively evenly distributed in numbers and volume, the results obtained are comparable with more quantitative methods (MacDonald & Green, 1983). A subjective estimate of gut fullness (on a scale from 0 = empty to 4 = full) was made on animals where the time interval between capture and dissection was known to be less than three hours. Guts from 296 goldsinny caught at Oban and Millport, 199 corkwing from Oban, and 30 rock cook from Oban and Millport were examined. Because of a time delay between capture and examination, no fish from Luing were suitable for dietary examination. It was not always possible to identify items to species level and hence some are given to only generic or family level.

2.2.4   *Condition indices*

Fish were killed by prolonged anaesthesia (ethyl p-amino benzoate; Sigma). All fish were measured to the nearest 0.5 mm SL and TL. After blotting dry on absorbent paper, the TW (total weight) of each specimen was obtained. The fish were dissected and the wet weights to the nearest 0.001 g of the gonads (GW), liver (LW) and guts obtained. After the organs were removed, the fish were weighed again to obtain the eviscerated weight (EW).

Gut weights were omitted from all calculations because bait was used in the capture of some of the fish. A somatic condition factor (K_S) was calculated as:

$$K_S = EW/aTL^b$$

the values of a and b being derived from the length-weight relationship EW = aTL^b for wrasse over one-year-old. To give an indication of the seasonal variation in the physical condition of the fish, the eviscerated weight (EW) was used to eliminate any influence of seasonal change in the weights of internal organs.

Gonadosomatic index (GSI) values were used in the present study as indicators of the degree of gonadal development, and were not intended to be accurate indicators of gonadal activity (DeVlaming *et al.*, 1982). Gonadal development was quantified by expressing gonad weight as a proportion of the eviscerated weight of the fish:

    GSI = 100 GW/EW

Variations in the LW were evaluated quantitatively as a hepatosomatic index (HSI):

    HSI = 100 LW/EW

Mean (±SE) K_S, GSI and HSI values were calculated at monthly intervals.

2.2.5   *Statistical analyses*

Examination of the condition indices by the correlation test for normality (essentially equivalent to the Shapiro-Wilk test) determined whether statistical analysis was by Student's *t*-test or the Mann-Whitney U-test (Zar, 1984).

2.3   **Results**

2.3.1   *Sex ratios and age structure*

Of the total of 475 goldsinny dissected for biological examination, 211 were male, 187 were female and 77 were juveniles that showed no signs of sexual differentiation. The overall sex ratio of 1:0.88 (male:female) was not significantly different from a 1:1 ratio (χ^2; $P > 0.05$). The sex ratios for goldsinny caught around Millport (1:1.32) and Oban (1:1.17) were also not significantly different from a sex ratio of 1:1 (χ^2; $P > 0.05$ in each case). The sex ratio for goldsinny caught around Luing (1:0.28) was significantly biased in favour of males (χ^2; $P < 0.001$). For age-groups 1, 2, 3 and 5 years, the sex ratio for goldsinny caught from Oban and Millport was not significantly different from 1:1 (χ^2; $P > 0.05$ in all cases).

    There were 74 male and 107 female corkwing of a total of 203; sexual identification was not possible for 22 specimens. The resultant sex ratio (1:1.45) was significantly different from a 1:1 ratio (χ^2; $P < 0.05$).

    Of the 63 rock cook dissected, 30 were male, 27 were female and 6 were of indeterminate sex; the overall sex ratio of (1:0.90) was not significantly different from a 1:1 ratio (χ^2; $P > 0.05$).

    For goldsinny, the age-groups 0–2 years were under-represented by creeling; 15% of the total number of goldsinny caught by creels were aged 0-2 years compared with 69% of those captured using quinaldine (Table 2.1). No similar sampling biases were recorded for corkwing or rock cook.

2.3.2   *Age and growth*

The use of sagittal otoliths for estimating age of the three wrasse species proved straightforward because each otolith had a white opaque nucleus surrounded by distinctive alternate transparent and opaque zones. A small number of otoliths were crystalline in appearance and did not have distinct banding. Rarely were

Table 2.1 Goldsinny catch-frequency data using baited creels or anaesthetic quinaldine (ND, sex not determined).

Age-group	Creels				Quinaldine				Totals		
	Male	Female	Total	%	Male	Female	Total	(%)	Male	Female	Total
0	0	0	0	0.0	ND	ND	77	45.8	ND	ND	77
1	6	1	7	2.3	5	8	13	7.7	11	9	20
2	22	17	39	12.7	11	15	26	15.5	33	32	65
3	65	47	112	36.5	10	7	17	10.1	75	54	129
4	38	21	59	19.2	3	4	7	4.2	41	25	66
5	23	15	38	12.4	6	6	12	7.1	29	21	50
6	5	4	9	2.9	0	2	2	1.2	5	6	11
7	4	9	13	4.2	1	3	4	2.4	5	12	17
8	3	9	12	3.9	1	0	1	0.6	4	9	13
9	2	2	4	1.3	0	1	1	0.6	2	3	5
10	3	1	4	1.3	1	1	2	1.2	4	2	6
11	1	0	1	0.3	0	5	5	3.0	1	5	6
12	0	1	1	0.3	0	0	0	0.0	0	1	1
13	0	1	1	0.3	0	0	0	0.0	0	1	1
14	1	0	1	0.3	0	0	0	0.0	1	0	1
15	0	1	1	0.3	0	1	1	0.6	0	2	2
16	0	1	1	0.3	0	0	0	0.0	0	1	1
17	0	1	1	0.3	0	0	0	0.0	0	1	1
18	0	0	0	0.0	0	0	0	0.0	0	0	0
19	0	2	2	0.7	0	0	0	0.0	0	2	2
20	0	1	1	0.3	0	0	0	0.0	0	0	0
Total	173	134	307		38	53	168		211	187	475

both otoliths from the same fish unreadable. The capture of early young-of-the-year fish allowed the otolith nucleus to be identified easily. Fish were caught throughout the year and it was therefore possible to follow the progression of otolith development. Otoliths examined from fish captured from late May to early December had a translucent edge to them; those from fish caught December through to May, an opaque edge.

Goldsinny

Male goldsinny of age one year and older ranged in size from 67–163 mm TL, weighed 3.75–64.10 g TW, and had a maximum age of 14+ years ($n = 211$). Females of one year and older ranged in size from 59–159 mm TL, weighed 2.41–65.37 g TW, and had a maximum age of 20+ years ($n = 185$). Juvenile goldsinny aged under one year ($n = 75$) were mostly of indeterminate sex, ranged in size from 16–57 mm TL and weighed 0.08–1.97 g TW.

For both sexes, there was no apparent difference in growth rates between fish caught from Oban and Luing; goldsinny from Millport were slower growing in

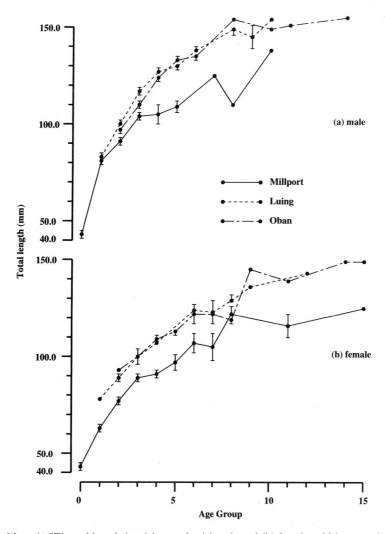

Fig. 2.2 Mean (±SE) total length (mm) by age for (a) male and (b) female goldsinny caught at three locations on the west coast of Scotland.

comparison (Fig. 2.2). Female goldsinny were slower growing compared with males at each of the three locations (Fig. 2.2).

Corkwing

Male corkwing of age one year and older ranged in size from 82–212 mm TL, weighed 7.1–131.2 g TW, and had a maximum age of 6+ years ($n = 73$). Females of one year and older ranged in size from 75–203 mm TL, weighed 5.5–103.5 g TW and had a maximum age of 7+ years ($n = 105$). Juvenile corkwing aged under one year ($n = 24$) were mostly of indeterminate sex, ranged in size from

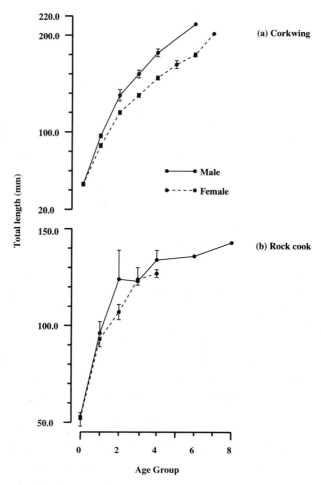

Fig. 2.3 Mean (±SE) total length (mm) by age for (a) corkwing and (b) rock cook males and females from the west coast of Scotland.

31–56 mm TL and weighed 0.3–1.8 g TW. Male corkwing grew faster than females (Fig. 2.3a).

Rock cook

Male rock cook of age one year and older ranged in size from 86–165 mm TL, weighed 6.8–56.5 g TW, and had a maximum age of 8+ years ($n = 30$). Females of one year and older ranged in size from 83–138 mm TL, weighed 7.9–39.3 g TW and had a maximum age of 5+ years ($n = 27$). Juvenile rock cook aged under one year ($n = 9$) were mostly of indeterminate sex, ranged in size from 43–71 mm TL and weighed 0.7–3.9 g TW. There was little difference in growth rates between sexes (Fig. 2.3b).

2.3.3 *Diet*

Goldsinny

The seasonal trend in fullness for goldsinny from Millport and Oban was similar, with greater fullness from June/July to November compared with the period December to May (Fig. 2.4). There were differences in the seasonal, annual and detailed composition of the diet between wrasse from Millport and Oban (Fig. 2.4, Table 2.2, and Table 2.3 respectively), and between sexes and different age-groups. Gastropod molluscs, amphipod crustaceans and polychaetes (predominantly serpulid worms) occurred in the diet of goldsinny from Millport in most months (Fig. 2.4a). However, in June/July bivalve molluscs (predominantly mussels) and barnacles (Cirripedia) were by far the dominant food items. Some guts contained over 100 mussels (2–4 mm in length). Food categories of less importance were hydrozoans, mysids (Mysidacea), copepods, decapods, isopods, ossicles from the tube feet of *Echinus esculentus* L., chironomid larvae (Insecta), and Rhodophyceae (Fig. 2.4a; Tables 2.2 and 2.3). Goldsinny aged zero plus, only captured from Millport, had a restricted diet, all containing amphipods, and over 75% had polychaete (predominantly serpulid) worms.

The dominant food categories from goldsinny captured at Oban were amphipods, barnacles, polychaetes (again predominantly serpulid worms), hydrozoans, and bivalve molluscs (predominantly mussels). All of these food groups were present in guts for most months when the fish were eating, apart from mussels which were eaten mostly from mid to late summer (Fig. 2.4b). The food categories of lesser importance were similar in occurrence to Millport goldsinny, apart from the absence of Mysidacea, Insecta and Rhodophyceae in the guts of Oban fish, and the presence of Pycnogonida and Teleostei (Table 2.2). The only consistent dietary difference between sexes was a tendency for female fish to consume more polychaetes at Oban and Millport, and Oban males to take more hydroids.

Corkwing

There was evidence of feeding throughout the year in both male and female adult corkwing (Fig. 2.5). Guts were at or over half-fullness from April to September; and below this level from October to March. There were few seasonal, annual or detailed differences between the diets of the two sexes (Fig. 2.5; Tables 2.4 and 2.5 respectively). The dominant food category for corkwing was gastropod molluscs [in particular *Gibbula umbilicalis* (da Costa) and *Helcion pellucidum* (L.)] both in terms of percentage occurrence (Table 2.4), and, subjectively, gut volume. Gastropods were present in corkwing gut samples for all months of the year. Categories of less importance were amphipods, ostracods, bivalve molluscs and polychaete worms. Bivalves were only eaten from March to November; there were no seasonal dietary trends for the other major foods.

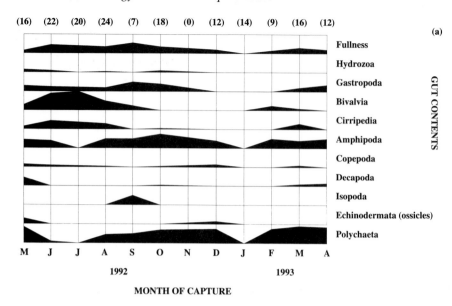

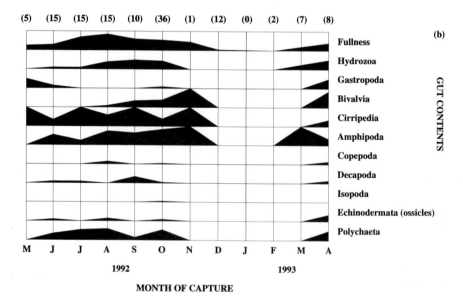

Fig. 2.4 Seasonal variation of some gut contents of male and female goldsinny caught at (a) Millport and (b) Oban. Fullness was subjectively rated on a scale of 0 (empty) to 4 (full) for guts dissected within 3 hours of the known time of capture; full bar = 4. Gut contents are quantified as the mean number of guts containing that food item expressed as a percentage of the total number of guts containing some food for the respective month; full bar = 100% (redrawn from Sayer *et al.*, 1995). Sample numbers in parentheses.

Table 2.2 Occurrence of food categories in guts of goldsinny collected from Millport and Oban (expressed as percentage of guts containing some food).

Location	Millport and Oban				Millport					Oban		
Sex	M, F, j¹	M, F	M	F	M, F,	M, F	M	F	j¹	M, F	M	F
Age	0+→	1+→	1+→	1+→	0+→	1+→	1+→	1+→	0+	2+→	2+→	2+→
Guts examined	296	252	101	151	170	126	46	80	44	126	55	71
Guts containing food	176	162	68	94	84	70	28	42	14	92	40	52
Gut contents												
HYDROZOA	21.0	22.8	25.0	21.3	5.9	7.1	0.0	11.9	0.0	34.8	42.5	28.8
MOLLUSCA												
Gastropoda	18.8	20.4	23.5	18.1	27.4	32.8	35.7	31.0	0.0	10.9	15.0	7.7
Bivalvia	26.7	29.0	30.9	27.7	28.6	34.3	32.1	35.7	0.0	25.0	30.0	21.2
CRUSTACEA												
Cirripedia	36.4	39.5	39.7	39.4	16.7	20.0	17.9	21.4	0.0	54.3	55.0	53.8
Amphipoda	58.0	54.3	57.4	52.1	51.2	41.4	42.9	40.5	100.0	64.1	67.5	61.5
Mysidacea	0.6	0.6	0.0	1.1	1.2	1.4	0.0	2.4	0.0	0.0	0.0	0.0
Copepoda	5.7	4.3	2.9	5.3	7.1	4.3	3.6	4.8	21.4	4.3	2.5	5.8
Decapoda	8.5	9.3	8.8	9.6	10.7	10.0	14.3	11.9	0.0	6.5	5.0	7.8
Isopoda	1.7	1.9	0.0	3.2	2.4	2.9	0.0	4.8	0.0	1.1	0.0	1.9
ECHINODERMATA												
ossicles	7.4	7.4	1.5	11.7	4.8	4.3	0.0	7.1	7.1	9.8	2.5	15.4
spines	4.0	4.3	4.4	4.3	6.0	7.1	3.6	9.5	0.0	2.2	5.0	0.0
POLYCHAETA	52.8	50.6	36.8	60.6	60.7	57.1	39.3	69.0	78.6	45.7	35.0	53.8
TELEOSTEI	0.6	0.6	1.5	0.0	0.0	0.0	0.0	0.0	0.0	1.1	2.5	0.0
PYCNOGONIDA	1.1	1.2	1.5	1.1	0.0	0.0	0.0	0.0	0.0	2.2	2.5	1.9
INSECTA	2.3	2.5	2.9	2.1	4.8	5.7	7.1	4.8	0.0	0.0	0.0	0.0
RHODOPHYCEAE	3.4	3.1	2.9	3.2	6.0	7.1	7.1	7.1	0.0	0.0	0.0	0.0

j¹ = most juvenile (0+) goldsinny examined were of indeterminate sex.

Table 2.3 Detailed composition of the diet of goldsinny caught at Millport and Oban ($\sqrt{}$ = present; $\times$ = absent).

Higher Taxa	Subordinate Taxa	Millport	Oban
HYDROZOA		$\sqrt{}$	$\sqrt{}$
MOLLUSCA	Gastropoda	*Helcion pellucidum* (L.)	*Helcion pellucidum* (L.)
		Littorina littorea (L.)	*Littorina littorea* (L.)
		Littorina 'littoralis' (L.)	*Littorina 'littoralis'* (L.)
		Patella vulgata L.	*Patella vulgata* L.
		Hydrobiidae	
		Crepidula fornicata (L.)	
		Nucella lapillus (L.)	
		Gibbula umbilicalis (da Costa)	
		Tectura virginea (Müller)	
		Eumarginula fissura L.	
		Buccinum undatum L.	
	Bivalvia	*Mytilus edulis* L.	*Mytilus edulis* L.
		Acanthocardia sp. Gray	*Venerupis* sp. Lamark
			Modiolus sp. Lamark
			Hiatella arctica (L.)
CRUSTACEA	Cirripedia	$\sqrt{}$	$\sqrt{}$
	Amphipoda	Gammaridae	Gammaridae
		Corophidae	Corophidae
	Mysidacea	$\sqrt{}$	$\times$
	Copepoda	Harpacticoida	Harpacticoida
		Metis ignea Philippi	
	Decapoda	*Galathea intermedia* Lilljeborg	*Galathea intermedia* Lilljeborg
		Galathea strigosa (L.)	*Galathea strigosa* (L.)
		Necora puber (L.)	*Pagurus bernhardus* (L.)
			Carcinus maenas (L.)
			Macropodia sp. Leach
	Isopoda	$\sqrt{}$	$\sqrt{}$
ECHINODERMATA		$\sqrt{}$	$\sqrt{}$
POLYCHAETA	Aphroditidae	*Harmothoë impar* (Johnston)	*Harmothoë impar* (Johnston)
	Spirorbidae	*Spirorbis spirorbis* (L.)	*Spirorbis spirorbis* (L.)
			Spirorbis corallinae da Silva and Knight-Jones
			Circeis spirillum (L.)
	Serpulidae	*Pomatoceros triqueter* (L.)	*Pomatoceros triqueter* (L.)
	Flabelligeridae	*Pherusa plumosa* (Müller)	
	Terebellidae	*Lanice conchilega* (Pallas)	*Lanice conchilega* (Pallas)
TELEOSTEI	Gobiesocidae		Clingfish
PYCNOGONIDA	Nymphonidae	$\times$	$\sqrt{}$
	Achelliidae	$\times$	$\sqrt{}$
INSECTA	Chironomidae	$\sqrt{}$	$\times$
RHODOPHYCEAE		$\sqrt{}$	$\times$

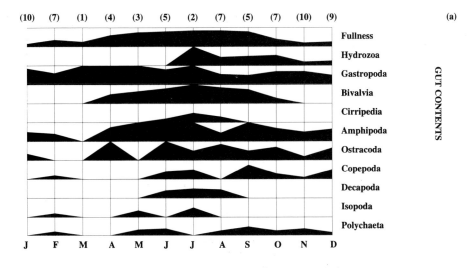

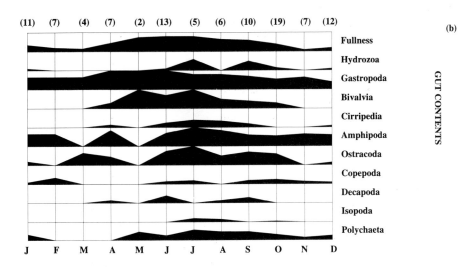

Fig. 2.5 Seasonal variation of some gut contents of (a) male and (b) female corkwing caught at Oban. Fullness was subjectively rated on a scale of 0 (empty) to 4 (full) for guts dissected within three hours of the known time of capture; full bar = 4. Gut contents are quantified as the mean number of guts containing that food item expressed as a percentage of the total number of guts containing some food for the respective month; full bar = 100%. Sample numbers in parentheses.

Table 2.4 Occurrence of food categories in guts of corkwing collected from around Oban (expressed as percentage of guts containing some food).

Sex	F	M	F + M	j^1
Age	$1+\rightarrow$	$1+\rightarrow$	$1+\rightarrow$	$0+\rightarrow$
Guts examined	103	75	178	21
Guts containing some food	94	62	156	18
Gut contents				
HYDROZOA	13.8	24.2	17.9	0.0
MOLLUSCA				
Gastropoda	71.3	71.0	71.1	5.6
Opisthobranchia	1.1	0.0	0.7	0.0
Bivalvia	31.9	33.9	32.7	0.0
Polyplacophora	0.0	1.6	0.6	0.0
CRUSTACEA				
Cirripedia	11.7	6.5	9.6	0.0
Amphipoda	63.8	64.5	64.1	22.2
Ostracoda	47.8	51.6	49.4	33.3
Copepoda	14.9	22.6	17.9	83.3
Decapoda	16.0	12.9	14.7	0.0
Mysidacea	1.1	0.0	0.7	0.0
Isopoda	3.2	4.8	3.8	0.0
ECHINODERMATA	6.4	3.2	5.1	0.0
POLYCHAETA	29.8	24.2	27.6	0.0
OLIGOCHAETA	0.0	1.6	0.6	0.0
TELEOSTEI				
Fish eggs	8.5	8.1	8.3	0.0
INSECTA	3.2	6.5	4.5	0.0
CHLOROPHYCEAE	1.1	3.2	1.9	0.0
RHODOPHYCEAE	3.2	3.2	3.2	0.0
SAND-GRAINS	24.5	17.7	21.8	0.0

Table 2.5 Detailed composition of the diet of male and female corkwing ($\checkmark$ = present; × = absent).

Higher Taxa	Subordinate Taxa	Male corkwing	Female corkwing
MOLLUSCA	Gastropoda	*Helcion pellucidum* (L.) *Littorina littorea* (L.) *Littorina 'littoralis'* (L.) Hydrobiidae *Gibbula umbilicalis* (da Costa) *Tectura virginea* (Müller)	*Helcion pellucidum* (L.) *Littorina littorea* (L.) *Littorina 'littoralis'* (L.) *Littorina 'saxatilis'* (Olivi) *Patella vulgata* L. Hydrobiidae *Gibbula umbilicalis* (da Costa)

Table 2.5 *Continued*

Higher Taxa	Subordinate Taxa	Male corkwing	Female corkwing
			Tectura virginea (Müller)
			Opisthobranchia
			Skeneopsis planorbis (Fabricus)
			Tricolia pullus (L.)
			Epitonium clathrus (L.)
	Bivalvia	*Mytilus edulis* L.	*Mytilus edulis* L.
		Musculus discors (L.)	*Venerupis* sp. Lamark
			Musculus discors (L.)
	Polyplacophora	*Lepidochitona cinereus* (L.)	
CRUSTACEA	Cirripedia	√	√
	Amphipoda	Gammaridae	Gammaridae
		Corophidae	Corophidae
	Ostracoda	√	√
	Mysidacea	×	√
	Copepoda	Harpacticoida	Harpacticoida
		Porcellidium viridae Philippi	
		Metis ignea Philippi	
	Decapoda	*Carcinus maenas* (L.)	*Carcinus maenas* (L.)
		Macropodia sp. Leach	*Macropodia* sp. Leach
		Hyas araneus (L.)	Palaemonidae
		Palaemonidae	Pandalidae
		Pandalidae	
	Isopoda	√	√
		Idotea neglecta Sars	
ECHINODERMATA		√	√
POLYCHAETA	Aphroditidae		*Harmothoë impar* (Johnston)
	Spirorbidae	*Spirorbis spirorbis* (L.)	*Spirorbis spirorbis* (L.)
			Spirorbis corallinae da Silva and Knight-Jones
	Phyllodocidae	*Phyllodoce laminosa* Savigny	
	Serpulidae		*Pomatoceros triqueter* (L.)
	Flabelligeridae	*Pherusa Plumosa* (Müller)	*Pherusa plumosa* (Müller)
	Terebellidae	*Lanice conchilega* (Pallas)	*Lanice conchilega* (Pallas)
		Polycirrus caliendrum Claparède	
OLIGOCHAETA	Tubificidae	*Tubificoides pseudogaster* (Dahl)	
INSECTA	Chironomidae	√	√
	Diptera	√	×
CHLOROPHYCEAE		√	√
RHODOPHYCEAE		√	√

Food items of minor importance for corkwing were hydrozoans, cirripeds, copepods, decapods, isopods, chlorophyceae, rhodophyceae, fish eggs and insects. Single opisthobranch, polyplacophoran, mysid, and oligochaete specimens were recorded. Sand-grains were recorded in over 20% of corkwing guts; sea urchin spines in over 5%.

Only four food items were found in the stomachs of young-of-the-year (<1 year of age) corkwing. The dominant food category was copepods found in 83% of guts with food. The two identified copepod species were the Harpacticoids *Metis ignea* Philippi and *Porcellidium viride* Philippi. Ostracods and amphipods were found in 33% and 22% respectively of juvenile guts containing some food, with bivalves in 6%.

Rock cook

Only 30 rock cook guts were available for contents analysis, and of that total only five male and ten female guts were found to contain any food items. Only four food items were found in the male stomachs: mussels (80% occurrence), gammarid amphipods (100%), harpacticoid copepods (60%), and an unidentified polychaete worm (20%). Female guts contained hydrobiid gastropods (20%), mussels (80%), a barnacle (10%), corophid (40%) and gammarid (40%) amphipods, harpacticoid copepods (30%) and ostracods (20%). No food items were found in any juvenile gut examined.

2.3.4   *Condition indices*

There were no significant relationships between K_S and TL, or between either GSI or HSI and EW ($P > 0.05$ in each case).

Goldsinny

K_S values for male and female goldsinny from Millport showed similar trends (Figs 2.6a and b) with values recorded from February to May significantly lower than at other times of the year ($P < 0.05$ in each case). Male and female goldsinny caught at Oban also had annual K_S trends that were similar (Figs 2.6c and d) but these trends were different from those recorded for Millport fish in that there were no significant increases in K_S during the summer months.

Over the whole year, mean ($\pm$SE) K_S values for Millport and Oban males (1.008 $\pm$ 0.007 and 1.010 $\pm$ 0.008 respectively) and females (0.995 $\pm$ 0.007 and 0.983 $\pm$ 0.009 respectively) were not different from each other intrasexually ($P > 0.05$ in both cases). Female K_S values were significantly lower than males in a combined intersexual comparison ($P < 0.05$).

Data from Luing goldsinny were insufficient to make any valid seasonal comparison. However, May to August K_S values for male and female goldsinny from

Luing (Table 2.6) were significantly lower than Millport values ($P < 0.01$ in both cases), and significantly lower than Oban values for females ($P < 0.01$).

GSI values for goldsinny collected from Millport reached their maxima in June (Figs 2.7a and b). The GSI peak was less evident in goldsinny from Oban, where

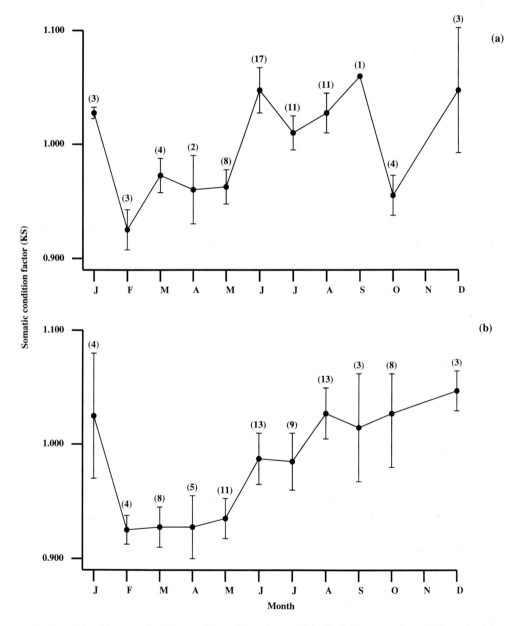

Fig. 2.6 Monthly mean (±SE) somatic condition factor (K_S) of goldsinny caught at Millport (male (a), female (b)) and Oban (male (c), female (d)) (redrawn from Sayer *et al.*, 1995). Sample numbers in parentheses.

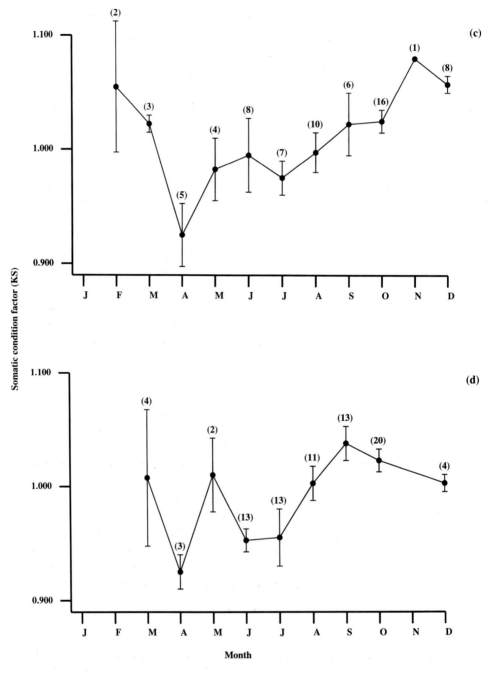

Fig. 2.6   *Continued*

Table 2.6 Monthly K_s, GSI and HSI values for male and female goldsinny (mean $\pm$ SE(n)) caught at Luing.

	May	June	July	August
Male goldsinny				
K_s	0.981 $\pm$ 0.023 (7)	1.002 $\pm$ 0.012 (30)	0.982 $\pm$ 0.010 (20)	0.993 $\pm$ 0.017 (15)
GSI	0.323 $\pm$ 0.097 (7)	0.287 $\pm$ 0.039 (28)	0.343 $\pm$ 0.080 (20)	0.176 $\pm$ 0.041 (15)
HSI	1.657 $\pm$ 0.269 (7)	1.631 $\pm$ 0.098 (30)	1.572 $\pm$ 0.100 (20)	1.953 $\pm$ 0.166 (15)
Female goldsinny				
K_s	0.962 $\pm$ 0.083 (3)	0.962 $\pm$ 0.011 (7)	0.938 $\pm$ 0.022 (8)	0.937 $\pm$ 0.049 (2)
GSI	2.400 $\pm$ 0.201 (3)	1.931 $\pm$ 0.236 (7)	4.851 $\pm$ 0.948 (8)	1.823 $\pm$ 0.139 (2)
HSI	1.428 $\pm$ 0.392 (3)	1.751 $\pm$ 0.221 (7)	1.793 $\pm$ 0.293 (8)	1.268 $\pm$ 0.269 (2)

May, June and July GSI values were all similar for both male and female (Figs 2.7c and d; $P < 0.05$ in all cases). The seasonal GSI trend for Luing males was similar to those from Oban with no significant peak from May to July ($P < 0.05$ in each case). There was a definite female GSI peak at Luing in July when values were significantly higher than June and August ($P < 0.01$ in both cases). For both sexes at all three locations the GSI had returned to predevelopment values by August.

Maximum GSI values for the three locations were 1.56, 1.24 and 1.64 for males (Millport, Oban and Luing respectively) and 17.41, 12.95 and 8.27 for females (Millport, Oban and Luing respectively). On average, peak GSI values for male and female goldsinny from Millport were significantly higher than those for the respective sexes at Oban and Luing ($P < 0.001$ in all cases), being approximately twice as high at Millport compared with Oban, and three times those at Luing (for both sexes). Maximum GSI values for both sexes at Oban and Luing were not significantly different from each other ($P > 0.05$ in both cases).

There were no obvious seasonal HSI trends for male goldsinny caught at Millport and Oban (Figs 2.8a and c). Female HSI values increased in the summer months at both Millport and Oban, but remained high at Oban for the autumn and early winter, whereas Millport values decreased (Figs 2.8b and d). No seasonal HSI trend was discernible from the limited data obtained from Luing.

Corkwing

Both male and female corkwing had significantly lower K_S values from April to July/August than at other times of the year ($P < 0.05$ in all cases; Figs 2.9a and b). Mean ($\pm$SE) K_S values for males (1.334 $\pm$ 0.011) and females (1.339 $\pm$ 0.009) over the whole year were not significantly different from each other ($P > 0.05$).

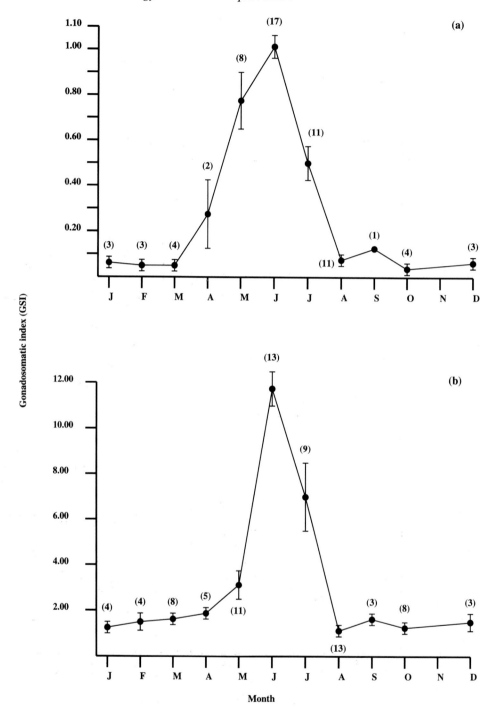

Fig. 2.7 Monthly mean (±SE) gonadosomatic index (GSI) of goldsinny caught at Millport (male (a), female (b)) and Oban (male (c), female (d)) (redrawn from Sayer *et al.*, 1995). Sample numbers in parentheses.

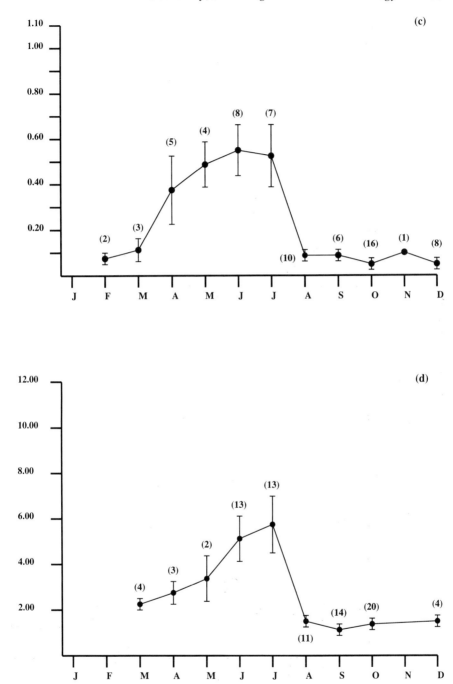

Fig. 2.7 *Continued*

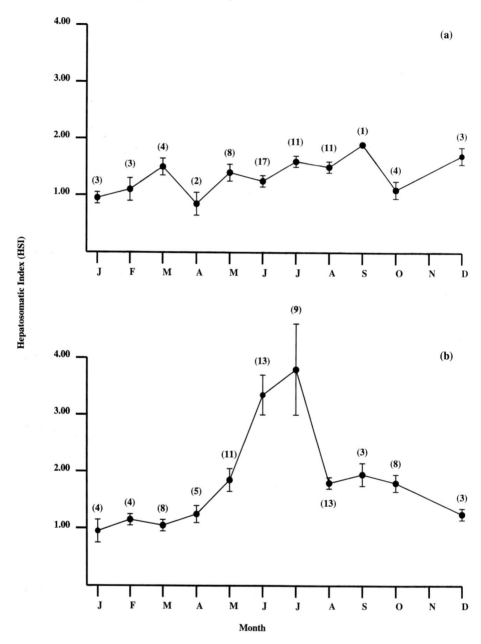

Fig. 2.8 Monthly mean (±SE) hepatosomatic index (HSI) of goldsinny caught at Millport (male (a), female (b)) and Oban (male (c), female (d)) (redrawn from Sayer *et al.*, 1995). Sample numbers in parentheses.

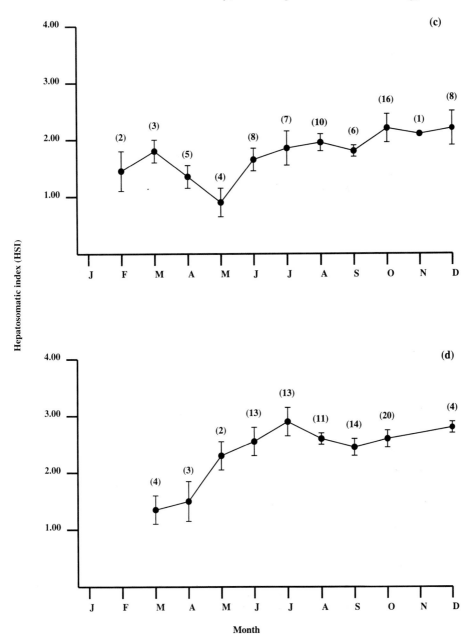

Fig. 2.8 *Continued*

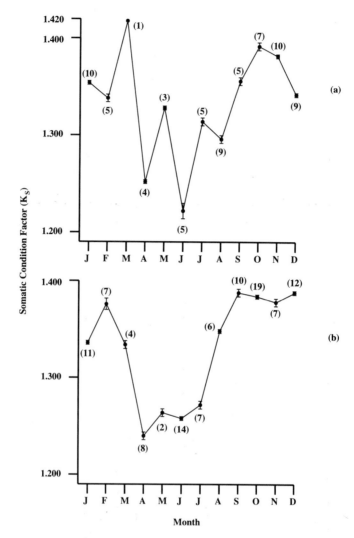

Fig. 2.9 Monthly mean (±SE) somatic condition factor (K_S) of (a) male and (b) female corkwing caught at Oban. Sample numbers in parentheses.

Gonadal development above autumn/winter levels was recorded from May to August (excluding September) for males, and April to August for females, with peak mean values both occurring in June (Figs 2.10a and b). Maximum GSI values for male and female corkwing were 11.37 and 18.94 respectively. There was no significant seasonal variation in male HSI values (Fig. 2.11a; $P > 0.05$ in all cases); female values for June to August were significantly higher than levels recorded over the rest of the year (Fig. 2.11b; $P < 0.001$).

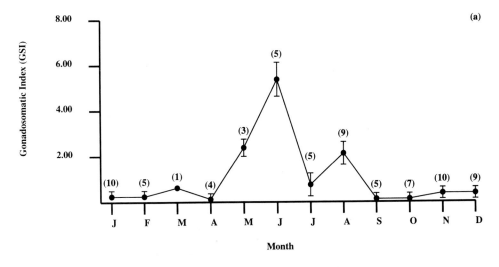

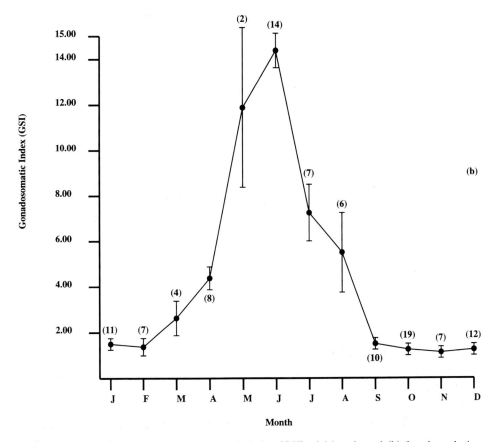

Fig. 2.10 Monthly mean (±SE) gonadosomatic index (GSI) of (a) male and (b) female corkwing caught at Oban. Sample numbers in parentheses.

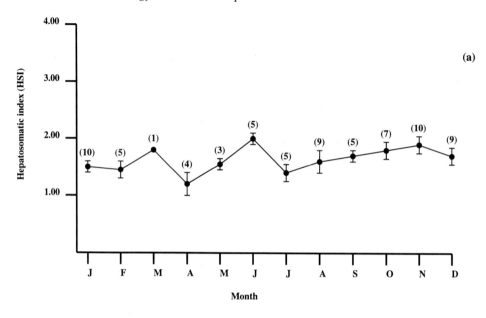

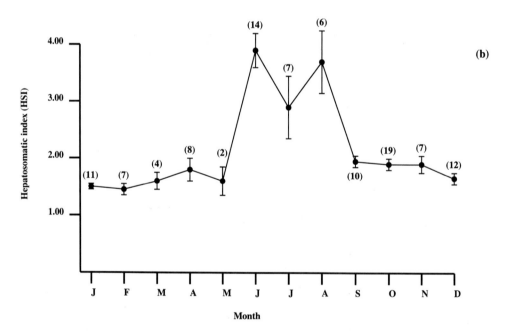

Fig. 2.11 Monthly mean (±SE) hepatosomatic index (HSI) of (a) male and (b) female corkwing caught at Oban. Sample numbers in parentheses.

Rock cook

K_S values for both male and female rock cook in August and September were higher than levels recorded at other times of the year (Figs 2.12a and b) although only significantly so for females ($P < 0.001$). Mean ($\pm$SE) K_S values for males (0.973 ± 0.021) and females (0.997 ± 0.023) over the whole year were not significantly different from each other ($P > 0.05$)

Even with the very limited data available, GSI levels for male and female rock

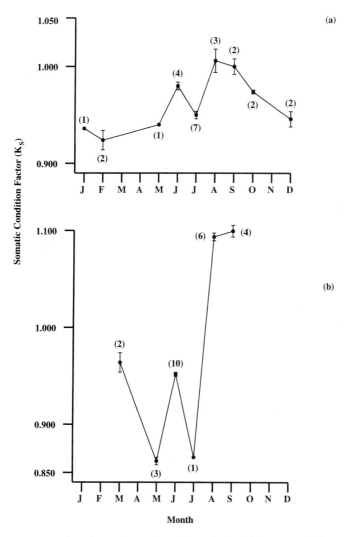

Fig. 2.12 Monthly mean ($\pm$SE) somatic condition factor (K_S) of (a) male and (b) female rock cook caught at Oban and Millport. Sample numbers in parentheses.

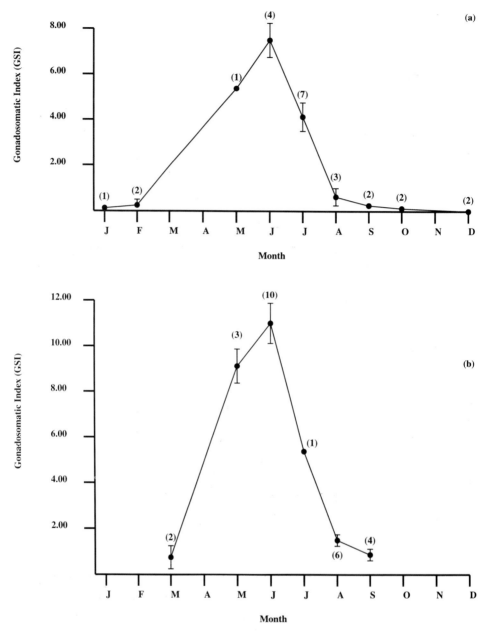

Fig. 2.13 Monthly mean (±SE) gonadosomatic index (GSI) of (a) male and (b) female rock cook caught at Oban and Millport. Sample numbers in parentheses.

cook each reached their maximum in June (Figs 2.13a and b respectively). Levels declined in July and were at baseline values by August. Data are lacking prior to the period of maximum GSI, but levels from May to July were significantly higher than values recorded during the rest of the year ($P < 0.001$ for both sexes). The

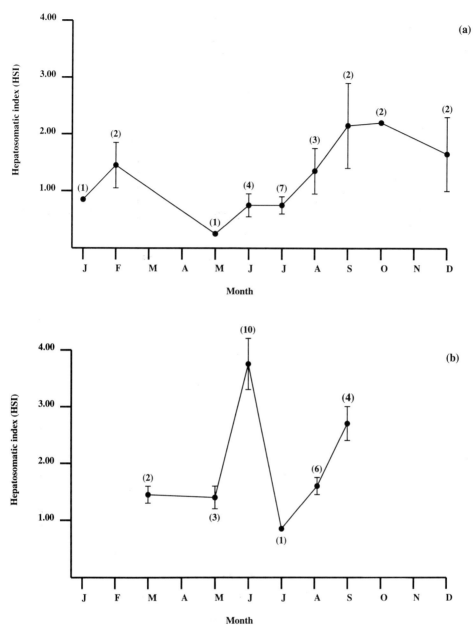

Fig. 2.14 Monthly mean (±SE) hepatosomatic index (HSI) of (a) male and (b) female rock cook caught at Oban and Millport. Sample numbers in parentheses.

maximum recorded GSI were 9.63 for males and 16.63 for females. During the period of elevated GSI values (May to July) male HSI levels were significantly lower than other values (Fig. 2.14a; $P < 0.05$); female HSI levels were significantly higher (Fig. 2.14b; $P < 0.05$).

2.4   Discussion

2.4.1   *Age and growth*

The maximum recorded ages for male and female wrasse in the present study were: 14+ and 20+ years respectively for goldsinny; 6+ and 7+ respectively for corkwing; and 8+ and 4+ respectively for rock cook. These estimates agree with those of Darwall *et al.* (1992) and Treasurer (1994) who gave maximum ages of 15, 9, 8, years and 16, 5, 9 years respectively. The annual incremental growth rates of all three wrasse species were highly variable, though sample numbers were low in some cases. Similar irregular growth was recorded for ballan wrasse, *Labrus bergylta* Ascanius (Dipper *et al.*, 1977). Allowing for this variation in growth pattern, there were clear locational and sexual differences in recorded growth rates. Males were longer-at-age than females for all three species at least from 1+ to 4+ years. A similar trend was reported by Treasurer (1994) who attributed slower female growth to earlier maturation. Although changes in growth could be caused in part by variation in the timing of sexual maturation, it is possible that differences in the relative scales of annual gonadal development may also be contributory factors.

Goldsinny caught at Millport grew slower than those captured at Oban and Luing. Growth rates of goldsinny caught on the mid-west coast of Scotland (Treasurer, 1994) were similar to those recorded in the present study for Millport. Rock cook caught in mid-west waters were slower growing compared with the present study; mid-west female corkwing growth was faster. Geographical differences in growth in a western Atlantic labrid [*Tautoga onitis* (L.)] were attributed to higher water temperatures at lower latitudes permitting growth over a longer season (Hostetter and Munroe, 1993). This does not appear to be the case with goldsinny, where fish from higher latitudes (Oban: this study; Sweden: Hilldén, 1978a; mid-west Scotland: Treasurer, 1994) grow at faster rates over the first four years compared with lower-latitude fish of the French Atlantic and Mediterranean coasts (Quignard, 1966), though the high variation between individual growth, and the low sample numbers for some age-groups may compromise such comparison. The disparity in growth between goldsinny from Millport and Oban cannot be explained by differences in seasonal water temperature regimes or duration of activity; water temperatures measured over one year were similar at the two locations (mean and range: Oban 10.9, 6.2–14.5°C; Millport 10.5, 5.8–14.2°C) and goldsinny at Millport were active longer (Sayer *et al.*, 1993). However, there were differences in the depth of water in which the fish were caught (3–8 m Millport, 12–25 m Oban), the density of goldsinny at the two sites (maxima: Oban 1.5 m^{-2}, Millport 4.0 m^{-2}; Sayer *et al.*, 1993), the diet and the magnitude of gonadal development (see section 2.4.3), any combination of which may account for the disparity in growth. Similar habitat or life history disparities may also explain different growth rates recorded for goldsinny, corkwing and rock cook between studies undertaken in different geographical locations. Geographical and sexual variation in growth rates has implications for wrasse fisheries where a lower limit of 10 cm total length is commonly implemented (this volume, Chapters 7, 8 and 9).

2.4.2   *Diet*

As in Sweden (Hilldén, 1978b), the dominant food categories of goldsinny were epibenthic organisms, notably Hydrozoa, Mollusca (Gastropoda and Bivalvia), Crustacea (Cirripedia and Amphipoda) and Polychaeta. The only exception was that bryozoans were found in relatively large quantities in Swedish goldsinny. There were also seasonal similarities between the two studies, showing a switch from feeding on polychaetes to bivalves during late spring/early summer when the newly-settled mussels were abundant and small enough to ingest.

The guts of newly-settled goldsinny contained predominantly amphipods copepods and polychaetes. Lebour (1920) also recorded amphipods, copepods and decapods in the guts of juvenile goldsinny. Harpacticoid copepods were predominant in the gut contents of juveniles in the present study, suggesting that they forage demersally at the post-settlement stage.

Compared with goldsinny, corkwing have a more predictable diet, with gastropod molluscs predominating. As with goldsinny, the diet can be described as epibenthic dependent. Because corkwing inhabit areas with macroalgal cover, they can graze both from the algal fronds and the sea bed. These feeding actions may explain, in part, the occurrence of Chlorophyceae, Rhodophyceae, sand grains and sea urchin spines, which could have been ingested coincidentally with the intended prey item. Dietary analysis of rock cook guts was relatively superficial because of low sample numbers. The gut contents that were examined were dominated by bivalve molluscs and amphipods, suggesting that rock cook were similar to goldsinny and corkwing in being epibenthic grazers. Because other studies of corkwing and rock cook diet are lacking, it cannot be assumed that the dietary analyses presented here are typical for the respective species. However, it is likely that feeding behaviour will be similar across areas even though dietary details differ.

A detailed knowledge of the diet of wrasse is of importance when assessing possible ecological effects of over-fishing. Wrasse are very numerous on the west coast of Scotland (Sayer *et al.*, 1993). Removal of large numbers may affect populations of both wrasse predators and prey. On a large scale, removal of a central predator and its prey may result in significant ecological alterations. A large proportion of the diets of goldsinny and corkwing wrasse are typical marine fouling organisms. There is an increasing volume of, as yet, anecdotal evidence to suggest that the inclusion of wrasse in salmon cages reduces the problem of net fouling. Systematic testing of their effectiveness is required; but even if the use of wrasse in controlling lice infestation becomes superseded by other methods, the continued employment of wrasse to clean nets remains a possibility.

2.4.3   Condition indices

There were no significant relationships between K_S and TL, or between either GSI or HSI and EW. Any trends in these indices are therefore unlikely to be artifacts of scaling. There is a tendency for seasonal changes in the magnitude of condition factors and HSI to be similar (Htun-Han, 1978; Tominaga *et al.*, 1991; Yokoyama *et al.*, 1991) possibly reflecting the utilization of the liver as a storage organ, the weight changes of which may mirror the overall somatic condition of the fish. Variations in liver weight reflect the process of storage and transfer of proteins and lipids associated with reproductive effort (Htun-Han, 1978; Crupkin *et al.*, 1988). It would therefore be expected that HSI values would be high during the pre-spawning period, and low during the spawning and immediate post-spawning periods. Similarly, changes in condition factor in several fishes have been ascribed to a depletion of body reserves during gonad maturation (Htun-Han, 1978). This relationship, whereby increases in GSI values are accompanied by a concomitant decrease in HSI and, in some cases, condition factor, has been reported for some fish species (Crupkin *et al.*, 1988; Tominaga *et al.*, 1991; Yokoyama *et al.*, 1991) but not others (Quignard, 1966; Htun-Han, 1978; Fujioka *et al.*, 1990).

In goldsinny, the pattern of gonad development was similar at both Millport and Oban. Male goldsinny GSI values began to increase in April and remained elevated until July; female GSI were elevated only for June and July. The main geographical difference was in the magnitude of increase in GSI, with the maximum GSI for goldsinny of each sex at Millport being twice the value of equivalent Oban maxima. Comparative data for goldsinny are not available, but compared with other labrid species maximum GSIs in male goldsinny were low, but female values were similar (goldsinny maxima: M 1.64, F 17.41, this study; other labrids: M 0.77–11.37, F 7.98–18.94, *Labrus merula* (L.), *L. viridis* (L.), *Crenilabrus tinca* (L.), *C. cinereus* (Bonnaterre), *C. mediterraneus* (L.), Quignard, 1966; *Tautoga onitis* , Hostetter and Munroe, 1993; *Crenilabrus melops*, *Centrolabrus exoletus*, this volume). Although there were differences in method of GSI calculation between the above studies, the use of EW will enhance GSI values in comparison with studies where TW was used. The higher level of investment in gonadal development by both male and female goldsinny from Millport may explain lower growth rates compared with fish from Oban. The period of increased gonad development for rock cook and corkwing was slightly more prolonged compared with goldsinny which may be reflective of the higher GSI values in those two species.

HSI values of male goldsinny and corkwing did not decrease markedly over the period of elevated GSI although HSI in females goldsinny, rock cook and corkwing increased. In both sexes, goldsinny at each location showed a tendency for their K_S values to be lower at the beginning of the reproductive period. In April/May, goldsinny activity increases markedly as water temperatures rise above 8°C

(Costello, 1991; Darwall *et al.*, 1992; Sayer *et al.*, 1993) and they also begin to feed again. Any depletion of liver reserves or reduction in somatic condition associated with gonadal development may be compensated for by increased food intake.

During the reproductive season goldsinny males invest considerable effort in patrolling/defending a territory whereas the females are less active (Sjölander *et al.*, 1972; Hilldén, 1981; Sayer, *pers. obs.*). Although the males may forage during this period (Hilldén, 1981) the proportion of time available to do so must be reduced. It might therefore be expected that male HSI and K_S would be low while the fish are highly territorial. This trend is not evident from the data presented here, but the comparatively low GSI values for male goldsinny compared with other labrids may be a result of the more intense territorial behaviour observed for goldsinny (Sjölander *et al.*, 1972; Hilldén, 1981; Sayer, *pers. obs.*). The energy available for gonad development may be reduced because of increased activity, although the investment required may be offset in part by having females protected in a harem. Circadian variation in HSI has been measured for rainbow trout (*Oncorhynchus mykiss* Walbaum; Boujard & Leatherland, 1992) and it is therefore possible that the monthly HSI estimates presented here are too crude to enable true trends to be identified.

Acknowledgements

Funding for this project was provided jointly by the Scottish Salmon Growers Association and the Crown Estate.

References

Bjordal, Å. (1991) Wrasse as cleaner-fish for farmed salmon. *Progress in Underwater Science* **16**: 17–28.

Boujard, T. and Leatherland, J.F. (1992) Circadian pattern of hepatosomatic index, liver glycogen and lipid content, plasma non-esterified fatty acid, glucose, T_3, T_4, growth hormone and cortisol concentrations in *Oncorhynchus mykiss* held under different photoperiod regimes and fed using demand-feeders. *Fish Physiology and Biochemistry* **10**: 111–12.

Costello, M.J. (1991) Review of the biology of wrasse (Labridae: Pisces) in northern Europe. *Progress in Underwater Science* **16**: 29–51.

Costello, M.J. (1993) Review of methods to control sea lice (Caligidae: Crustacea) infestations on salmon (*Salmo salar*) farms. In *Pathogens of Wild and Farmed Fish Sea Lice* (Ed. by Boxshall, G.A. and Defaye, D.). Ellis Horwood, Chichester, pp. 219–52.

Crupkin, M., Montecchia, C.L. and Trucco, R.E. (1988) Seasonal variations in gonadosomatic index, liver-somatic index and myosin/actin ratio in actomyosin of mature hake (*Merluccius hubbsi*). *Comparative Biochemistry and Physiology* **89A**: 7–10.

Darwall, W.R.T., Costello, M.J., Donnelly, R. and Lysaght, S. (1992) Implications of life-history strategies for a new wrasse fishery. *Journal of Fish Biology* **41** (Supplement B), 111–23.

DeVlaming, V., Grossman, G. and Chapman, F. (1982) On the use of the gonadosomatic index. *Comparative Biochemistry and Physiology* **73A**: 31–9.

Dipper, F.A., Bridges, C.R. and Menz, A. (1977) Age, growth and feeding in the ballan wrasse *Labrus bergylta* Ascanius 1767. *Journal of Fish Biology* **11**: 105–20.

Fujioka, T., Takahashi, T., Maeda, T., Nakatani, T. and Matsushima, H. (1990) Annual life cycle and distribution of adult gurnard *Lepidotrigla microptera* in Mutsu Bay, Aomori Prefecture. *Bulletin of the Japanese Society of Scientific Fisheries* **56**: 1553–60.

Hilldén, N.-O. (1978a) An age-length key for *Ctenolabrus rupestris* (L.) in Swedish waters. *Journal du Conseil International pour l'Exploration de la Mer* **38**: 270–1.

Hilldén, N.-O. (1978b) On the feeding of the goldsinny, *Ctenolabrus ruprestris* L. (Pisces, Labridae). *Ophelia* **17**: 195–8.

Hilldén, N.-O. (1981) Territoriality and reproductive behaviour in the goldsinny, *Ctenolabrus rupestris* L. *Behavioural Processes* **6**: 207–21.

Hostetter, E.B. and Munroe, T.A. (1993) Age, growth, and reproduction of tautog *Tautoga onitis* (Labridae: Perciformes) from coastal waters of Virginia. *Fishery Bulletin, US* **91**: 45–64.

Htun-Han, M. (1978) The reproductive biology of the dab *Limanda limanda* (L.) in the North Sea: gonadosomatic index, hepatosomatic index and condition factor. *Journal of Fish Biology* **13**: 369–78.

Hyslop, E.J. (1980) Stomach contents analysis–a review of methods and their application. *Journal of Fish Biology* **17**: 411–29.

Lebour, M.V. (1920) The food of young fish, No. III (1919). *Journal of the Marine Biological Association of the United Kingdom* **12**: 261–324.

MacDonald, J.S. and Green, R.H. (1983) Redundancy of variables used to describe importance of prey species in fish diets. *Canadian Journal of Fisheries and Aquatic Sciences* **40**: 635–7.

Qasim, S.Z. (1957) The biology of *Blennius pholis* L. (Telestei). *Proceedings of the Zoological Society of London* **128**: 161–208.

Quignard, J.-P. (1966) Recherches sur les Labridae (Poissons, Teleostéens, Perciformes) des côtes européennes: systématique et biologie. *Naturalia Monspeliensia (Zoologie)* **5**: 7–248.

Sayer, M.D.J., Cameron, K.S. and Wilkinson, G. (1994) Fish species found in the rocky sublittoral during winter months as revealed by the underwater application of the anaesthetic quinaldine. *Journal of Fish Biology* **44**: 351–3.

Sayer, M.D.J., Gibson, R.N. and Atkinson, R.J.A. (1993) Distribution and density of populations of goldsinny wrasse (*Ctenolabrus rupestris*) on the west coast of Scotland. *Journal of Fish Biology* **43** (Supplement A): 157–67.

Sayer, M.D.J., Gibson, R.N. and Atkinson, R.J.A. (1995) Growth, diet and condition of goldsinny on the west coast of Scotland. *Journal of Fish Biology* **46**: 317–40.

Sjölander, S., Larson, H.O. and Engström, J. (1972) On the reproductive behaviour of two labrid fishes, the ballan wrasse (*Labrus bergylta*) and Jago's goldsinny (*Ctenolabrus rupestris*). *Revue Comporte Animale* **6**: 43–51.

Tominaga, O., Nashida, K., Maeda, T., Takahashi, T. and Kato, K. (1991) Annual life cycle and distribution of adult righteye flounder *Pleuronectes herzensteini* in the coastal waters of northern Niigata Prefecture. *Bulletin of the Japanese Society of Scientific Fisheries* **57**: 2023–31.

Treasurer, J.W. (1993) Management of sea lice (Caligidae) with wrasse (Labridae) on Atlantic salmon (*Salmo salar* L.) farms. In *Pathogens of Wild and Farmed Fish Sea Lice* (Ed. by Boxshall, G.A. and Defaye, D.). Ellis Horwood, Chichester, pp. 335–45.

Treasurer, J.W. (1994) The distribution, age and growth of wrasse (Labridae) in inshore waters of west Scotland. *Journal of Fish Biology* **44**: 905–18.

Yokoyama, S., Maeda, T., Takahashi, T., Nakatani, T. and Matsushima, H. (1991) Annual life cycle of adult *Hippoglossoides dubius* in Funka Bay, Hokkaido. *Bulletin of the Japanese Society of Scientific Fisheries* **57**: 1469–76.

Zar, J.H. (1984) *Biostatistical Analysis* (2nd edn). New Jersey, Prentice Hall.

Chapter 3
Observations of wrasse on an artificial reef

K.J. COLLINS, A.C. JENSEN and J.J. MALLINSON *Department of Oceanography, University of Southampton, Southampton SO17 1BJ, UK*

A small experimental artificial reef was constructed in Poole Bay, on the central south coast of England, in June 1989, to study the environmental compatibility of stabilized coal ash and its fishery-enhancement potential. Colonization by fish, crustacea and epibiota was rapid. Wrasse species were observed within the first month. Reef block settlement by epifauna and flora reached almost 100% coverage during the first summer. After four years the species composition and density has become comparable with those of neighbouring natural reefs. Wrasse species (corkwing, *Crenilabrus melops*; ballan, *Labrus bergylta* and goldsinny, *Ctenolabrus rupestris*) form the main part of the resident fish population. These have been observed feeding on the reef biota. Each May/June from 1990 to 1994, male corkwing wrasse have been recorded building nests on each of the eight reef units. Drift and reef algae were used to construct a two layer structure. Normally, several nests were started on each unit, though typically only one per unit was completed and maintained. Nests were guarded and tended for three to four weeks whilst the eggs hatched. Time-lapse video was used to record courtship and diurnal behaviour on and around the nest.

The Poole Bay study has demonstrated the potential for creating new wrasse habitat, providing niches for reproduction and feeding. Reefs could be built in association with marine fish farms to act as biofilters (to remove excess nutrients), sites for wrasse production and possibly lobster farming.

3.1   Introduction

A small experimental artificial reef was constructed in Poole Bay, on the central south coast of England, in June 1989 to study the environmental compatibility of stabilized coal ash and its fishery-enhancement potential (Collins *et al.*, 1991). Colonization of the reef block surfaces by epibiota and utilization by mobile fauna, fish and crustacea are important indicators of the value and development of such a structure. Pelagic fish are attracted to a wide variety of marine structures, sometimes without gaining significant benefit from them. This has led to the debate as to whether artificial reefs actually promote marine productivity, or simply attract existing stocks which can then be more readily exploited (Seaman

& Sprangue, 1991). Territorial species, such as wrasse, present a clearer case for a positive contribution to the marine ecosystem, as they use reefs for shelter from currents and predators, and for feeding and reproduction, during much of their life cycle.

The commercial exploitation of wrasse species for control of sea lice infestation of farmed salmon described by other papers in this volume has led to concern about the sustainability of natural populations (this volume, Chapter 8). The colonization of a new artificial habitat has parallels with the recolonization of a heavily exploited natural habitat. In both cases the rate of occupation provides a measure of the mobility of surrounding adult stocks and the transport of planktonic larvae. This study describes the colonization of an artificial reef by wrasse species. Detailed observation of one wrasse species (corkwing) was undertaken to demonstrate the value of the artificial reef to the species (in particular and fishery enhancement in general) during part of its life cycle.

3.2   Methods

The reef consists of eight units in two rows with centres 10 m apart. Each unit is a conical pile (1 m high by 4 m dia.) of 6 tonnes of randomly stacked stabilized coal ash blocks (20 × 20 × 40 cm). This random arrangement provides a large surface area and a variety of crevices and galleries within the reef units. About 25% of the volume of each unit is void space. The reef lies in 10 m (below chart datum) on a sandy seabed, over 3 km from any rocky seabed habitat.

The artificial reef was studied intensively by scientific divers. The frequency of diving was monthly, increasing to weekly during the summer months. Divers quantified fish numbers by swimming two transects, one over and the second around the perimeter of each reef unit.

Corkwing wrasse behaviour was filmed with a Sony Hi8 V900 video camera in an Amphibico housing with 100W filming lamp mounted on a tripod 0.5–1 m from the nest recording continuously for up to one hour. Longer-term studies were made with a time lapse system comprising a Sony V90 video camera in custom housing containing controlling circuitry and power supply for a 50W lamp which was switched on when filming during the night. The camera was positioned 60 cm away from corkwing nests with a field of view approximately 60 cm across. Five seconds of film was recorded every two-and-a-half minutes over a period of 22 hours.

3.3   Results

3.3.1   *Reef colonization*

Colonization of the reef was rapid. Pouting [*Trisopterus luscus* (L.)] shoaled around the structure from the day of deployment. This has remained the dominant fish species, with shoals of over 100 individuals observed around each reef unit during the daytime in summer. Wrasse species were observed within the first month. Lobsters and crabs also moved in during this time and subsequently have been intensively studied (Jensen *et al.*, 1994).

Hydroids, ascidians and red algae dominated cover of the block surfaces during the first summer. In subsequent years bryozoans and sponges have increased in abundance. There is a clear annual cycle in epibiota abundance dominated by small red algae (typically 10 cm high) which reaches a maximum in late summer. Silt on the seabed of Poole Bay is readily resuspended by storms leading to limited underwater visibility (0–5 m) for much of the year. Consequently kelp (*Laminaria* spp.) growth is not found at the reef depth (10 m) in Poole Bay. After five years reef-block epibiota species composition and density has become comparable with that of neighbouring natural reefs.

3.3.2   *Wrasse observations*

Since the reef is over 3 km away from natural rocky habitat, the time of arrival of wrasse species is of interest. The initial colonizers were all adult fish. The numbers of wrasse [corkwing, *Crenilabrus melops* (L.); ballan, *Labrus bergylta* Ascanius and goldsinny, *Ctenolabrus rupestris* (L.)] on each reef unit were calculated from direct observations by divers in August 1994. Two independent counts of wrasse on each of the eight reef units were made and the results averaged (Table 3.1). Most of the ballan wrasse were around 20 cm long, though most reef units appeared to support a larger (30–40 cm) specimen. Corkwing, goldsinny and ballan juveniles (~40 mm) were observed during this survey. Two other wrasse species have been observed in Poole Bay, [cuckoo, *Labrus mixtus* L., and rock cook, *Centrolabrus exoletus* (L.)], but only individual specimens of cuckoo wrasse have been occasionally sighted on the reef during the five years since deployment.

Corkwing and goldsinny wrasse were very closely associated with the reef units staying close to, or within the structure. Ballan wrasse tended to be more mobile

Table 3.1 Dates of the first observed presence of wrasse species on the artificial reef (deployed 13.6.89), and average numbers of fish per reef unit counted by divers, August 1994.

Wrasse species	Date of first sighting	Days elapsed after reef deployment	Number per reef unit
Corkwing	5.7.89	22	2.5
Ballan	12.7.89	29	3.6
Goldsinny	19.7.89	46	2.3
Cuckoo	18.8.89	66	infrequent

and were observed swimming from one reef unit to the next. They were often seen lying against single blocks between reef units or feeding on the adjacent seabed which supports large numbers of slipper limpet [*Crepidula fornicata* (L.)] chains and their associated epifauna.

Cleaning behaviour by goldsinny wrasse was observed on red mullet (*Mullus surmuletus* L.) around the base of reef units and on ballan wrasse on a natural reef nearby.

Corkwing wrasse have been observed feeding on the reef epifauna. Patches of barnacles have been eaten; sponges and ascidians have also shown evidence of bite marks. Analysis of corkwing wrasse (liver and tissue), representing a higher trophic level, has shown no excess heavy metal concentrations over samples from control sites (Collins & Jensen, 1995).

3.3.3 Corkwing wrasse nesting

In June 1990, one year after deployment, brightly coloured male corkwing wrasse were first observed building seaweed nests between reef blocks. At this stage reef algal growth was limited and the fish were seen to collect passing drift fronds from the water column. This continued in subsequent years, starting in May 1991, 1992 and 1993; June in 1990. In 1994, diving on the reef did not start until June, missing the beginning of nest construction. Each year, several nests were con-structed on each reef unit although many were abandoned. This is illustrated by the sequence of events in 1991 (Fig. 3.1) which shows the location of nests observed from May to July. Many fish started construction of individual nests on each reef unit which appeared to be abandoned before completion. Typically only one large (>30 cm across) nest structure was completed per reef unit and then constantly attended for three to four weeks. The period (May to July) during

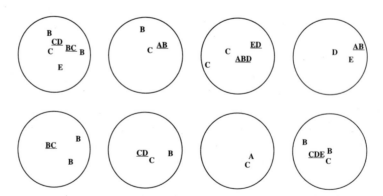

Fig. 3.1 Locations of corkwing wrasse nests, during May to July 1991, on the eight artificial reef units (not drawn to scale–reef units 4 m across, 6 m apart). Key: A = 16.5.91, B = 25.5.91, C = 21.6.91, D = 11.7.91, E = 23.7.91; nests recorded at the same position on more than one occasion are underlined.

MONTH		May				June				July				August		
WEEK	1	2	3	4	1	2	3	4	1	2	3	4	1	2	3	4
YEAR																
1990						X		X	X		X					
1991			X	X			X	X		X					X	
1992	X	X			X		X									
1993		X					X									
1994						X	X	X								

Fig. 3.2 Time-span of corkwing wrasse nesting activity (construction and attendance) on the artificial reef, showing the weeks of each month during which observations were made (**X**).

which nest building and attendance was noted is shown in Fig. 3.2. The male wrasse were very protective of their nests, chasing away other fish, and were not intimidated by the presence of divers.

During the day the activity of male corkwing appeared to be continuous, divided between rearranging the nest material and patrolling within about 2 m of the nest. Nest material was collected from reef blocks or drift weed floating by. Mouthfuls of weed were pushed into the nest with a prolonged wriggling action of the body. When examined, the structure was found to be complex, reminiscent of a bird's nest, with two distinct layers. The under-layer was based on coarse brown algae [*Sargassum muticum* (Yendo) Fensholt, and *Cystoseria* spp.]; and red algae [*Calleblepharis jubata* (Gooden and Woodward) Kützing, and *Cystoclonium purpureum* (Hudson) Batters]; whilst the upper, outer visible region was composed of softer mainly red species, predominantly *Plocamium cartilagineum* (L.) Dixon (Table 3.2). The extensive use of drift material was confirmed by the presence of many species not observed growing on the artificial reef. A few foliose and erect hydroids and bryozoans were included in the nests. Material was also removed from the nest and discarded up to 2 m away.

The activity of one animal was successfully recorded using time-lapse video throughout a day-night cycle (15–16.6.94). Excursions from the nest became less frequent towards dusk and ceased soon after nightfall, at 20:40 hrs BST. During the night the fish lay on one side at the base of the nest, occasionally changing position. On one occasion (00:49 hrs) it appeared to be ventilating the nest with a finning action. The fish started moving about at 04:30 and resumed excursions at 05:09 just before dawn.

Another behaviour revealed by the time-lapse filming but not previously observed by the divers on the reef was the interaction with a female. During the course of several hours the pair were seen circling the nest and approaching closely, though entering the nest was not recorded.

Table 3.2   Comparison of the species composition of the inner and outer layers of a single wrasse nest from the artificial reef (p = present, <1%).

Higher Taxa	Species	Inner % mass	Outer % mass
Angiospermae	*Zostera* sp.		p
Chlorophyceae	*Enteromorpha* sp.	p	p
	Chaetomorpha sp.	p	p
	Ulva sp.		p
Phaeophyceae	*Sargassum muticum* (Yendo) Fensholt	15	1
	Cystoseira nodicaulis (Withering) Roberts	11	
	Cystoseira baccata (Gmelin) Silva	10	
	Desmarestia viridis (O.F. Müller) Lamouroux	3	
	Cystoseira tamariscifolia (Hudson) Papenfuss	1	
	Halidrys siliquosa (L.) Lyngbye	p	
	Fucus sp.	p	
	Dictyota dichotoma (Hudson) Lamouroux	p	
Rhodophyceae	*Calliblepharis jubata* (Gooden & Woodward) Kützing	11	1
	Cystoclonium purpureum (Hudson) Batters	10	11
	Gracilaria bursa-pastoris (Gmelin) Silva	10	1
	Laurencia obtusa (Hudson) Lamouroux	8	4
	Halopitys incurvus (Hudson) Batters	7	1
	Ceramium sp.	4	3
	Plocamium cartilagineum (L.) Dixon	4	59
	Gracilaria sp.	1	
	Polyides rotundus (Hudson) Greville	1	
	Calliblepharis ciliata (Hudson) Kützing		3
	Phyllophora pseudoceranoides (Gmelin) Newroth & Taylor	p	
	Phyllophora crispa (Hudson) Dixon	p	
	Griffithsia flosculosa (Ellis) Batters		8
	Heterosiphonia plumosa (Ellis) Batters		5
	Hypoglossum woodwardii	p	p
	Champia parvula (C. Agardh) Harvey	p	
	Rhodophyllis sp.	p	p
	Corallina officinalis L.	p	p
	Apoglossum ruscifolium (Turner) J.G. Agardh		p
	Cryptopleura ramosa (Hudson) Newton		p
	Rhodomela sp.		p
Hydrozoa	*Hydrallmania falcata* (L.)	p	
	Sertularia argentea L.	p	
	Sertularella fusiformis (Hincks)	p	
	Obelia longissima (Pallas)	p	
Bryozoa	*Bowerbankia gracillima* (Hincks)	p	
	Bugula simplex Hincks	p	
	Flustra foliacea (L.)	p	

3.4   Discussion

The reef was designed primarily as a material test to determine the environmental impact of introducing stabilized coal ash into the marine environment. Block samples and epibiota have been analysed to determine if there was any evidence for leaching and consequent bio-accumulation of heavy metals in excess of that found naturally. No significant leaching or uptake has been found. After five years reef block epibiota species composition and density has become comparable with that of neighbouring natural reefs.

The colonization of the reef, 3 km distant from rocky habitat, soon after deployment is indicative of the mobility of adult wrasse species. Whilst goldsinny and ballan wrasse tend to be only associated with rocky habitat, corkwing wrasse have occasionally been found over sandy seabed (Sayer, *pers. comm.*). Wrasse species are territorial (Darwall *et al.*, 1992) but presumably there is some fraction of a population that will leave a site to find new habitat. Juvenile specimens of corkwing, goldsinny and ballan have been observed, suggesting the possibility of on-site recruitment for all three species, though only corkwing nesting has been observed.

Cuckoo wrasse, only occasionally sighted on the reef, are more commonly observed along the coast to the west of Poole Bay. The coast becomes more rocky with underwater rock ledges and boulder slopes, and water clarity improves away from the sedimentary bays such as Poole Bay and others to the east. These factors permit the growth of kelps possibly providing a more suitable habitat for the cuckoo wrasse. The absence, to date, of rock cook from the artificial reef is more surprising since these are commonly seen on natural patch reefs within Poole Bay. This suggests that the artificial reef is not within the normal range of adult movement.

Potts (1974) described the bright colouration and aggressive territorial behaviour of male corkwing wrasse. Aggressive behaviour is shown towards males and females, but behaviour towards females changes when the nest is ready to receive eggs. Courtship and spawning behaviour has been noted in detail by Potts (1974). Only courtship was recorded by the time lapse video in the present study. In an examination of corkwing nest structure from shallow sites off Wembury in Devon, Potts (1985) described a two-layer structure comparable to that found on the artificial reef. In both cases the inner layer was composed of coarser algal species. The main differences are in the species used in construction, reflecting the different depths (littoral, Wembury; 10 m, Poole Bay). Potts' observations of the great care with which material is selected and placed many times before final incorporation into the nest have been confirmed by the video recordings on the artificial reef.

Artificial reefs are frequently constructed with the aim of enhancing fisheries. The small-scale experimental Poole Bay artificial reef has provided habitat, feeding and reproduction opportunities for wrasse species and other commercial animals such as lobster. Another potential role for artificial reefs is the creation of a biofilter. The provision of hard substrate colonized by algae and filter feeders, such as mussels, near a marine fish farm may have the capacity to remove significant quantities of suspended solids and nutrients from the water column. In a UK context, it seems feasible for an artificial reef to be built in association with salmon farming, providing: populations of wrasse for cleaning, which could be returned to the reef to overwinter, (see this volume, Chapter 22); biofiltration to maintain water quality; and a possible by-catch of lobsters.

References

Collins, K.J., Jensen, A.C. and Lockwood, A.P.M. (1991) Artificial Reefs: using coal fired power station wastes constructively for fishery enhancement. *Oceanologica Acta* **11**: 225–9.

Collins, K.J. and Jensen, A.C. (1995) Stabilised coal ash artificial reef studies. *Chemistry and Ecology* **10**: 193–203.

Darwall, W.R.T., Costello, M.J., Donnelly, R. and Lysaght, S. (1992) Implications of life history strategies for a new wrasse fishery. *Journal of Fish Biology* **41** (Supplement B): 111–23.

Jensen, A.C., Collins, K.J., Free, E.K. and Bannister, R.C.A. (1994) Lobster (*Homarus gammarus*) movement on an artificial reef; the potential use of artificial reefs for stock enhancement. *Crustaceana* **68**: 198–211.

Potts, G.W. (1974) The colouration and its behavioural significance in the corkwing wrasse, *Crenilabrus melops*. *Journal of the Marine Biological Association of the United Kingdom* **54**: 925–38.

Potts, G.W. (1985) The nest structure of the corkwing wrasse, *Crenilabrus melops*, (Labridae: Teleostei). *Journal of the Marine Biological Association of the United Kingdom* **65**: 531–46.

Seaman, W. and Sprague, L.M. (eds) (1991) *Artificial Habitats for Marine and Freshwater Fisheries*. Academic Press Inc., San Diego.

Chapter 4
Field observations of the cleaning of ballan wrasse and torment feeding

M. BREEN *SOAFD Marine Laboratory, P.O. Box 101, Victoria Road, Aberdeen AB9 8DB, UK*

Field observations were made by SCUBA diving on the behaviour of ballan wrasse (*Labrus bergylta*) in Lough Hyne marine nature reserve, south-west Ireland. Cleaning symbiosis was recorded between ballan (as the host) and rock cook (*Centrolabrus exoletus*), the cleaner, on ten separate occasions. Ballan were observed to invite cleaning by rock cook, but attempts by goldsinny (*Ctenolabrus rupestris*) to clean were resisted. Ballan were observed to ingest food particles ejected into the water column by edible crabs (*Cancer pagurus*) as they burrowed into sediment following attacks by the ballan.

4.1   Introduction

Wrasse (*Labridae*) are a common and widespread family in both temperate and tropical waters (Nelson, 1994). The family is characterized by their thick lips, heavy teeth and bright, often brilliant coloration (Wheeler, 1969; Dipper, 1987; Lythgoe & Lythgoe, 1991). Ballan (*Labrus bergylta* Ascanius) appear to be the most abundant and widespread of the wrasse species inhabiting north European waters (Couch, 1868; Wheeler, 1969; Dipper, 1987), though no quantitative distributional information is available (Costello, 1991). They are most often found near rocky cliffs and reefs down to depths of 20–30 m (Lythgoe & Lythgoe, 1991), and feed mainly on molluscs (Quignard, 1966; Costello, 1991; Lythgoe & Lythgoe, 1991).

  Cleaning symbiosis is a mutually beneficial behaviour resulting in the removal of unwanted and/or harmful materials from the host organism while simultaneously furnishing food for the cleaner (Feder, 1966). Cleaning behaviour has been noted for all five species of wrasse that are common in northern Europe: ballan; goldsinny, *Ctenolabrus rupestris* (L.); rock cook, *Centrolabrus exoletus* (L.); corkwing, *Crenilabrus melops* (L.); and cuckoo, *Labrus mixtus* L. (Costello, 1991). This behaviour has only been recorded for juvenile ballan, however, though adults have previously been observed being cleaned by corkwing and goldsinny in aquaria, and rock cook in the field (Potts, 1973). This brief report gives a more detailed account than that given by Potts (1973) of adult ballan being cleaned by rock cook in the field. In addition, observations of previously unreported ballan feeding behaviour are described.

4.2   Materials and methods

Observations of ballan wrasse behaviour were made in Lough Hyne marine nature reserve in south-west Ireland (50°29′N 9°18′W) between 22 July and 1 September 1989. An area of sediment and rock-scree slope, approximately 55 m wide and covering a depth range of 0–18 m, was marked out with three transects at 8, 12 and 16 m depth. Behaviour was recorded by the same diver at six different times of day: 'dawn'; 08:00; 11:00; 14:00; 17:00; and 'dusk'. Each dive consisted of a gradual ascent from 18 m following the same route across the study area. Individual ballan were chosen at random and observed for an approximately four-minute period, during which time their behaviour was noted at five-second intervals. Observations were only made in low-water current conditions.

4.3   Results

4.3.1   *Cleaning symbiosis*

A symbiotic cleaning relationship was observed between ballan and rock cook. Ballan would approach individuals or groups of rock cook on the scree slope. This approach was made slowly and was followed by a period of hovering in close proximity to the rock cook, in an attempt to invite the rock cook to inspect. Ballan would normally adopt a horizontal orientation while hovering, though vertical orientations were also noted. If interested, individual rock cook would approach the ballan and slowly inspect it. Usually, rock cook would approach the caudal fin of the host first, if the ballan was hovering horizontally, or the ventral side if the ballan was hovering vertically. While inspecting ballan, rock cook would move slowly along the fish maintaining a constant orientation with the mouth 3–5 cm from and perpendicular to the body surface. Rock cook would stop and quickly remove any parasites or tissue of interest before returning to its usual orientation relative to the ballan (Fig. 4.1). No physical contact between ballan and rock cook was observed, other than the act of cleaning. The end of the cleaning period was marked by the slow retreat of rock cook at which point the ballan would turn and swim off. Cleaning behaviour was observed on ten occasions over the six-week period. It was not noted what the rock cook were removing from the ballan and no individual rock cook were removed for stomach content analysis.

While being inspected, the ballan would hover, maintaining its position with slow finning movements of its pectoral fins. It was noted that while hovering vertically, the ballan would stop finning with one pectoral to allow the close inspection by rock cook of the base of that pectoral fin and regions of the ventral side adjacent to it. During the inspection, the ballan would carefully watch the progress of the cleaner.

Associated with this symbiotic cleaning behaviour, were sporadic bouts of hovering by ballan followed by rapid scratching of their ventral surfaces against

Fig. 4.1   A ballan wrasse being cleaned by a rock cook.

underwater rocks. Prior to cleaning, individual ballan were often to be seen hovering and scratching, after which the fish would actively pursue rock cook. This behaviour was also seen on a number of occasions after ballan had apparently failed to attract rock cook to inspect or clean.

Goldsinny were observed to mimic rock cook cleaning behaviour and would approach and clean a ballan while rock cook were inspecting and cleaning. On the occasions that this was observed to occur, the ballan chased the goldsinny away. Rock cook were also observed attempting to clean pollack, *Pollachius pollachius* (L.), and smaller individuals of ballan. In both cases the potential hosts appeared irritated by the presence of the cleaner and actively avoided its often persistent attentions.

4.3.2   *Torment Feeding*

Ballan were observed hovering, at a height of *c.* 50–100 cm, above edible crabs, *Cancer pagurus* L., which were partially buried in shallow depressions in the sediment. The ballan would dive down and strike the crab on the carapace, causing the crab to burrow frantically further into the substratum. As a result, a large amount of the surrounding sediment and its contents were disturbed (Fig. 4.2). The ballan would then feed on items which had been released into the water column by the movement of the crab. This was a relatively minor activity, never seen at dawn or dusk, but infrequently between 08:00–17:00, with a peak occurrence at 14:00 hours.

Fig. 4.2   *Cancer pagarus* burrowing into sediment following an attack from a ballan wrasse.

4.4   Discussion

4.4.1   *Symbiotic Cleaning*

In the present study, rock cook, and to a lesser extent goldsinny, were observed attempting to clean other fish, predominantly ballan. Corkwing were not observed cleaning, as they were of low abundance in the study area.

In a comparison of the cleaning behaviours of four wrasse (goldsinny, rock cook, cuckoo and ballan), Bjordal (1988) found rock cook to be by far the most aggressive cleaners (ballan showed no cleaning ability). This could explain why only rock cook were observed cleaning successfully, but fails to explain why goldsinny were avoided by ballan. It is possible that being cleaned by rock cook is more beneficial to ballan. Continued advances by goldsinny suggest they may be opportunistic cleaners as far as ballan are concerned, although more accepted cleaning symbioses may exist with other species. As both rock cook and goldsinny are used in large numbers in salmon farming, an investigation as to whether similar preferences are exhibited by salmon, and the relative benefits of the respective wrasse species, may be of value.

Rock cook were not observed to perform elaborate, ritualized displays, characteristic of tropical cleaning wrasse (Eibl-Eibesfeldt, 1955; Randall, 1958, 1962; Wickler, 1963; Losey, 1971) which use displays to attract host fish. A lack of display by cleaners, with the host instead attracting the cleaners using its own displays (as shown by ballan in the present study), is considered characteristic of temperate cleaning symbioses (Potts, 1973). It is possible that this is indicative of the cleaning relationship at a lower evolutionary level than those of tropical symbioses and that temperate cleaners are not dependent on attracting host fish to them (Potts, 1973). Furthermore, Potts (1973) suggested that the invitation postures displayed by the host have evolved from comfort-seeking movements, such as bracing and chafing, which are orientated towards specific objects. When chafing, the ballan rubbed specific parts of their bodies against a conspicuous abrasive surface which may have relieved them of irritation. The removal of such an irritation could provide the stimulus which would reinforce the host in learning its role in a symbiotic cleaning relationship (Potts, 1973). This hypothesis is supported by the observations in the present study, where ballan displayed both invitation postures and chafing behaviour. Chafing behaviour was often observed immediately prior to invitation postures from ballan, indicating possible epidermal irritation.

Congregations of hosts and cleaners have previously been observed in distinctive areas ('cleaning stations') usually marked by a conspicuous physical feature (Randall, 1958; Limbaugh, 1961; Feder, 1966; Youngbluth, 1968). No cleaning stations were observed in the present study, though individual ballan were observed to travel to groups of rock cook.

4.4.2   Torment Feeding

The digging activities of edible crabs result in the excavation of interstitial species from the sediments, making them available as food for the ballan. Hall *et al.* (1992), noted marked increases in feeding activities of juvenile gadiods around crab pits within one minute of their creation. Plumes of sediment produced by the normal burrowing activities of crabs could act as a stimulus for ballan to approach the crab pits in order to feed. Ballan were attracted by a diver artificially disturbing the sediment.

The reason for the ballan attacking edible crabs is uncertain. The attack could be a predatory one directed at the crab itself. Although ballan may be incapable of breaking the carapace of a mature edible crab, a recently-moulted crab could provide easy prey. The initial attack may therefore be a means of testing the carapace hardness. The sediment and interstitial fauna released into the water column by the crab burrowing after the attack may be of secondary importance as a potential food source to the ballan. However, it cannot be dismissed that tormenting crabs, in order to feed on debris thrown up by their escape response, may be the primary reason for such behaviour by ballan for a significant proportion of attacks.

References

Bjordal, Å. (1988) Cleaning symbiosis between wrasses (Labridae) and lice infested salmon (Salmo salar) in mariculture. *International Council for the Exploration of the Sea, Mariculture Committee* 1988/F **17**: 8 pp.

Costello, M.J. (1991) Review of the biology of wrasse (Labridae: Pisces) in northern Europe. *Progress in Underwater Science* **16**: 29–51.

Couch, J. (1868) *A History of the Fishes of the British Islands* Vol. III. Groombridge and Sons, London.

Dipper, F. (1987) *British Sea Fishes*. University World Publications Ltd, London.

Eibl-Eibesfeldt, I. (1955) Über Symbiosen, Parasitismus und andere besondere zwischenartlicke Besiehungen tropischer Meeresfiscke. *Zeitschrift für Tierpsychologie, Berlin* **12**: 203–19.

Feder, H.M. (1966) Cleaning symbiosis in the marine environment. In *Symbiosis, I* (Ed. by Henry, S.M.). Academic Press, New York, pp. 327–80.

Hall, S.J., Basford, D.J., Robertson, M.R., Raffaelli, D.G. and Tuck, I. (1991) Patterns of recolonisation and the importance of pit-digging by the crab *Cancer pagurus* in a subtidal sand habitat. *Marine Ecology Progress Series* **72**: 93–102.

Limbaugh, C. (1961) Cleaning symbiosis. *Scientific American* **205**: 42–9.

Losey, G.S. (1971) Communication between fishes in cleaning symbioses. In *Aspects of the Biology of Symbiosis* (Ed. by Cheng, T.C.). Butterworth, London, pp. 45–76.

Lythgoe, J.N. and Lythgoe, G.I. (1991) *Fishes of the Sea: The North Atlantic and Mediterranean*. Blandford Press, London.

Nelson, J.S. (1994) *Fishes of the World*, 3rd edn. John Wiley and Sons, Inc., New York.

Potts, G.W. (1973) Cleaning symbiosis among British fish with special reference to *Crenilabrus melops* (Labridae). *Journal of the Marine Biological Association of the United Kingdom* **53**: 279–93.

Quignard, J.-P. (1966) Recherches sur les Labridae (Poissons, Teleostéens, Perciformes) des côtes européennes: systématique et biologie. *Naturalia Monspeliensia (Zoologie)* **5**: 7–248.

Randall, J.E. (1958) A review of the labrid fish genus *Labroides*, with descriptions of two new species and notes on ecology. *Pacific Science* **12**: 327–47.

Randall, J.E. (1962) Fish service stations. *Sea Frontiers* **8**: 40–7.

Wheeler, A. (1969) *Fishes of the British Isles and North-East Atlantic*. Macmillan, London.

Wickler, W. (1963) Zum Problem der Signalbildung, am Beispiel der Verhaltens Mimikry zwischen *Aspidontus* und *Labroides* (Pisces, Acanthopterygii). *Zeitschrift für Tierpsychologie, Berlin* **20**: 657–79.

Youngbluth, M.J. (1968) Aspects of the ecology and ethology of the cleaning fish, *Labroides phthirophagus* Randall. *Zeitschrift für Tierpsychologie, Berlin* **25**: 915–32.

Chapter 5
The territorial range of goldsinny wrasse on a small natural reef

K.J. COLLINS *Department of Oceanography, University of Southampton, Southampton SO17 1BJ, UK*

The population of goldsinny wrasse of part $(100\,\mathrm{m}^2)$ of a small natural reef in Poole Bay, on the British south coast was studied by SCUBA diving. Animals were caught in baited traps, marked with streamer tags and released. Population estimates using the Petersen and Schnabel census methods were in the order of 0.8 goldsinny m^{-2}. Observation of the tagged fish around the trap site suggested a range of movement or territory size for single fish some 10 m across. Combining this territory size with the population estimates a density of one fish m^{-2} was indicated.

A simple, iterative spreadsheet model was devised which allocated each fish a territory based on random movement within a defined radius about a home base/centre. Using a density of one fish m^{-2}, a radius of movement of 5 m and attraction to bait up to 2.5 m away, results comparable to the observed densities and recapture rates were obtained.

5.1   Introduction

Goldsinny wrasse were examined on a small rocky reef in Poole Bay, on the central south coast of England. The substratum of the Bay is largely composed of sand and silt with patch reefs (including Poole Rocks), typically 100–500 m across. The rocks are formed of collapsed sedimentary sandstone/ironstone strata, and provide a complex habitat with crevices and galleries at many levels. These are fished commercially for lobsters [*Homarus gammarus* (L.)] and edible crabs (*Cancer pagurus* L.). Tagging studies of the lobster population have been undertaken from 1990 to 1994 (Collins *et al.*, 1993). One reef to the south west of Poole Rocks has been studied in detail (lobsters and epibiota) as a natural control site for comparison with an artificial reef 3 km to the south. The numbers of wrasse, in particular large ballan wrasse (*Labrus bergylta* Ascanius), led to it being referred to as 'Wrasse Reef'. The reef is crescent shaped, 150 m long by 20 m wide, consisting of a pile of rocks ranging in size from 20 cm to over 1 m across. In common with other reefs in the area, there is a multilayered structure providing numerous crevices. Maximum height is about 2 m above the surrounding seabed 10–11 m below chart datum. Tidal currents reach 0.5 m/sec^{-1} during spring tides.

The area is silty, reducing light penetration, and limiting algal growth to small red species on the upper surfaces of larger rocks.

Being large (30–50 cm), ballan wrasse were the most conspicuous fish, but the most numerous appeared to be the goldsinny wrasse [*Ctenolabrus rupestris* (L.)], which were attracted to the bait used when lobster pots were set. Rock cook [*Centrolabrus exoletus* (L.)], of similar size to the goldsinny, occur at approximately one-tenth the density. Other cryptic fish species include tompot blennies [*Parablennius gattorugine* (L.)], black gobies (*Gobius niger* L.), and rock gobies (*G. paganellus* L.). Large grey mullet–either *Chelon labrosus* (Risso) or *Liza ramada* (Risso)–have been observed sheltering from tidal currents between the larger boulders.

Wrasse species are territorial (Darwall *et al.*, 1992) exploiting the shelter from tidal currents and predators afforded by the rocky habitats that they occupy. Those that guard nests/eggs ensure greater hatching success. This study aimed to determine the density and range of movement of goldsinny on part of Wrasse Reef.

5.2   Methods

Density of goldsinny was initially estimated by SCUBA divers counting the fish seen in a 1 m-wide measured transect across the study area. The study was undertaken during August 1994.

Goldsinny were trapped and released at one marked location on the southern arm of the reef. Two cylindrical polypropylene mesh traps with conical entrances were used baited with fish [pouting, *Trisopterus luscus* (L.)]. The traps measured 30 cm wide by 55 cm long (5 mm mesh) and 45 cm by 90 cm long (10 mm mesh) respectively. These were placed by divers and left for 40 mins before being closed and then hauled slowly to the surface, at a rate of 10 m in 20 mins. Trapped fish were anaesthetized with MS-222 (tricaine methanesulphonate) dissolved in seawater (50–100 mg l^{-1}). Numbered yellow, 50 mm, streamer tags (Hallprint Pty, Australia) were inserted between the dorsal fin bones into the anterior musculature. These were large for the 10–15 cm fish but had the advantage that the numbers could be read by divers. After 30 mins recovery in fresh seawater the fish were returned to the capture location by divers.

Distribution of the tagged fish around the trap site was observed by divers. North-south and east-west transect lines, marked at 5 and 10 m intervals from the centre, were laid through the trap site. At each marked point bait (pouting) was removed from a plastic bag to prevent leaving a scent trail and secured to a rock. The numbers and species of fish attracted to an area 1 m around the bait was noted over a ten-minute period. The transect lines were then rotated through 45° degrees and the exercise repeated.

5.3   **Results**

Of the two fish traps, the larger, bigger mesh trap was more successful. Table 5.1 shows the total numbers of fish caught by both traps plus the number tagged and subsequently recaptured. Traps were placed pointing downwards at an angle of about 30° from horizontal and the bait hung in a bag inside the trap close to the entrance funnel. Two simple population estimate techniques were applied to the tagging data; the Petersen single census and Schnabel multiple census for closed populations (Seber, 1982). The results of calculations (Table 5.2) over three successive recapture days indicate a population in the order of 80 goldsinny (0.8 fish m^{-2}).

The complex structure of the reef provided much shelter for the fish and therefore direct visual census was difficult. Observed goldsinny density was about 0.5 fish m^{-2}. An active approach was to use bait to attract fish from the crevices. The numbers of goldsinny observed at stations 5, 10, 15 and 20 m from the trap site are shown in Fig. 5.1. Very few fish were attracted onto the open seabed, composed of silty sand, beyond the edge of the reef. On the south east tip of the reef the continuous rock coverage changed to isolated boulders. The record at

Table 5.1 Numbers of trapped, tagged and recaptured goldsinny wrasse.

	Date			
	13.8.94 (day 1)	15.8.94 (day 2)	20.8.94 (day 3)	27.8.94 (day 4)
Total caught	16	18	12	30
Number tagged	13	16	6	10
Recaptured from day 1		2	3	2
Recaptured from day 2			1	6
Recaptured from day 3				2

Table 5.2 Population estimates made from the mark/recapture results in Table 5.1.

Method	Data set	Population (fish m^{-2})
Petersen single census	day 2	0.89
	day 3	0.70
	day 4	1.01
Schnabel multiple census	day 2	0.78
	day 3	0.83
	day 4	0.86

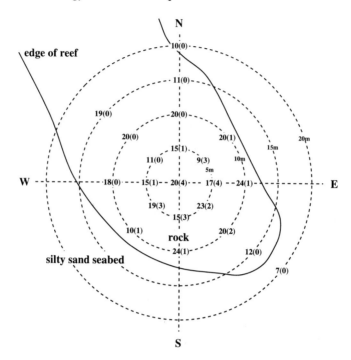

Fig. 5.1 Numbers of goldsinny wrasse attracted to bait during a 10-minute period at points around the trap/release site on a small natural rocky reef. Numbers of tagged fish are shown in brackets.

20 m from the trap site was obtained from bait set on one of these small rocks. The tag numbers on all the marked fish were recorded to determine if there was movement between the stations (Fig. 5.2). The distribution of the tagged fish indicated distinct overlapping ranges/territories. This led to the concept of a home base and a circle of activity about this point. The minimum radius of such a circle to encompass the observed movements was estimated to be in the order of 5 m. Of 35 fish tagged, 26 were subsequently observed during dives in the following 4 weeks. Five additional fish were seen to have lost tags, accounting for 89% of the released animals.

Spreadsheets provide a convenient and direct way of keeping data on a large number of individuals, manipulating and tracking their positions. In this case, Borland Quattro Pro spreadsheet PC software was used to model the movements of individual fish.

5.4   Discussion

5.4.1   *Population calculations*

Since the study was short term and the species is known to be territorial (Costello, 1991; Darwall *et al.*, 1992) closed system population estimates (Petersen single

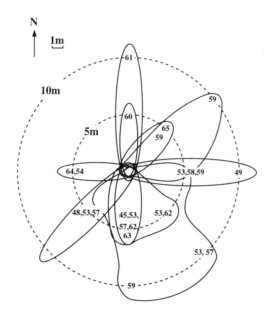

Fig. 5.2 Tag numbers of marked goldsinny wrasse observed around the trap/release site, indicating the ranges of individuals.

census and Schnabel multiple census; Seber, 1982) were considered adequate to provide a first approximation to the abundance of the fish.

5.4.2   *Spreadsheet model*

Given a static population moving within definite territories then the trap census must be drawing on only those fish whose ranges encompass the trap site, i.e. in the order of 5 m. Table 5.2 indicates a population of about 80 fish, which if based within a circle of 5 m radius ($79 \, m^2$) would give a density of one $fish \, m^{-2}$. It is not surprising that this is greater than the directly observed density ($0.5 \, fish \, m^{-2}$) given the opportunities for concealment afforded by the habitat. These figures are comparable to estimates by Sayer *et al.* (1993) who quote maximum estimates for three west of Scotland sites of 3.9, 1.3 and 1.1 $fish \, m^{-1}$. Observed numbers of fish attracted to bait were typically around 20, one quarter that of the calculated population above, suggesting that these are drawn from an area of approximately $20 \, m^2$ (radius 2.5 m).

The spreadsheet was used to model the movements of individual fish with home bases randomly distrbuted within a $30 \times 30 \, m$ area centred on the trap site and extending beyond the observed spread of tagged fish. The density of fish was changed by varying the numbers of animals used in the model (e.g. 900 fish/$900 \, m^2$ = 1 $fish \, m^{-2}$).

A radius of movement of 5 m about each home base was used initially. The position of an animal was calculated from two random functions (distance and bearing from grid zero). A simple random distance calculation generated equal probability of the animal being at any point along the radius thus the generated ground density was inversely related to radius. Any fish which was within a specified distance (initially 2.5 m) of the bait was assumed to be attracted towards it (Fig. 5.3). Attraction to the bait distance was set at half the home range radius. The spreadsheet formulae are given in Table 5.3.

The spreadsheet was set to manual recalculation. When this was commanded, the random functions generated a complete new data set. The initial occurrence of animals at the trap site was recorded permanently in column J and taken to represent the tagged population. This was set to 25 tagged fish to simulate the actual numbers (29 fish originally tagged, with 4 lost tags) present at the time of the observations (20.8.94). The occurrence of a fish at a point was recorded as a '1' for present and '0' for absent. Logical comparison with column J identified the tagged individuals. Summing columns yielded the total numbers of individuals. A data set from running/recalculating the model eight times to simulate the observed data is given (Table 5.4). The model is iterative, thus density, home-range radius and attractive range of bait were changed by altering the appropriate values in the spreadsheet and recalculating.

Results comparable to the observed fish numbers were obtained using a lower density (0.75 fish m^{-2}) with larger home range (6 m) and bait attraction range (3 m) or using higher density (2 fish m^{-2}) with smaller home range (4 m) (Table 5.4.) The closeness of the modelled (number of fish seen and those tagged) to the

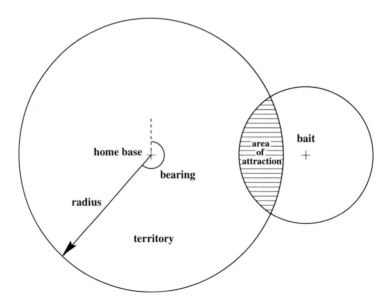

Fig. 5.3 Diagram to illustrate the concepts and calculations made in the spreadsheet model.

Table 5.3 Formulae used in the spreadsheet (Borland Quattro Pro) model (n = row number = an individual fish, the home base x and y coordinates are generated as random functions then converted to their fixed values before subsequent runs/recalculations).

Column	Description	Formula
A	home base x coordinate	@RAND*30-15
B	home base y coordinate	@RAND*30-15
C	distance from home base	@RAND*5
D	bearing	@RAND*2*@PI
E	x coordinate fish	+An+@SIN(Dn)*Cn
F	y coordinate fish	+Bn+@COS(Dn)*Cn
G	distance from centre grid/trap site	@SQRT(En*En+Fn*Fn)
H	distance of fish from station 5 m from trap site	@SQRT((En*En)+(Fn-5)(Fn-5))
I	distance fish from station 10 m from trap site	@SQRT((En*En)+(Fn-10)(Fn-10))
J	tagged fish = those attracted run 1	value Kn, run 1
K	attracted to bait at trap site	@IF (Gn<2.5,1,0)
L	repeat observation tagged animal at trap site	@IF(Kn=1)#AND#(Jn=1,1,0)
M	fish attracted to 5 m station	@(IF(Hn<2.5,1,0)
N	tagged fish seen at 5 m station	@IF(Jn=1)#AND#(Mn=1,1,0)
O	fish attracted to 10 m station	@IF(In<2.5,1,0)
P	tagged fish seen at 10 m station	@IF(In=1)#AND#(On=1,1,0)

Table 5.4 Comparison of observed numbers of goldsinny wrasse (total and tagged, see Fig. 5.1) with results from the spreadsheet model using different densities and home base radii. The closeness of fit between the models and observed numbers is indicated in each case by a Student's t-test, t value: r = home base radii (m); d = goldsinny density (number m^{-2}).

Distance (m)	Observed		Model, d = 1, r = 5		Model, d = 2, r = 4		Model, d = 0.75, r = 6	
	seen	tagged	seen	tagged	seen	tagged	seen	tagged
0	20	4	19	4	22	7	22	1
5	15	1	14	1	24	5	17	2
5	9	3	19	3	24	1	18	4
5	17	4	19	1	20	1	20	3
5	23	2	17	2	19	2	23	1
5	15	3	25	2	22	1	23	1
5	19	3	14	1	21	1	16	5
5	15	1	22	2	21	1	24	0
5	11	0	16	2	25	3	21	0
10	20	0	19	0	25	0	21	0
10	20	1	16	0	23	0	22	0
10	24	1	15	0	22	0	21	0
10	20	2	15	1	21	0	20	1
10	24	1	22	1	16	0	17	0
10	10	1	23	0	23	0	25	0
10	18	0	13	0	23	0	17	0
10	20	0	15	2	25	0	17	0
t			0.125	0.675	3.565	0.501	1.984	0.304

observed data was tested using Student's *t*-test (two-sample, with equal variances tool, Borland Quattro Pro). A density of one fish m^{-2} provided the closest approximation to the total numbers of fish observed. The home range of 4 m required by the higher density model is too small to enable sufficient tagged fish to be seen at 10 m from the origin. There is scope to manipulate the density of fish, distance of attraction to the bait, radius and distribution of movement about the home base.

The range of movement within a circle of some 5 m radius which overlaps the range of other goldsinnies contrasts with the findings of Hilldén (1981). This Swedish study on rocky seabed between 0.5 m and 10 m deep found a mean territory size of 1.4 m^2. Hilldén found male goldsinny to occupy a central hole and patrol the boundary, defending the territory against other males, unripe females, subadults and rock cook. Females stayed within the territory of the male that they had spawned with. Macroalgae, *Laminaria* and *Fucus* spp. were a significant feature of the study site used by Hilldén (1981) and when an area was cleared of algae it was not occupied by any territorial males. At the site used in the present study, there was very little algae, with occurrence restricted to small filamentous red species on the tops of boulders. The algae were unlikely to be large enough to afford any cover to the goldsinny. Hilldén (1981) showed that territory defence continued beyond the May to June period of spawning through to October. This would suggest that such behaviour would be expected during the current August/September study.

A possible explanation for the marked difference in behaviour is that the goldsinny population in the present study may be composed of immature fish. Preliminary examination of the otoliths from one goldsinny specimen of 10 cm total length indicated that it was less than one year-old. This implies much higher growth rates than those found in populations on the west coast of Scotland (Sayer *et al.*, 1995). However, Sayer *et al.* (1995) report large differences in individual growth rates within goldsinny populations and it is therefore not possible to make any inference about population growth rates from the age at length relationship of one individual. Further work on the age structure of the goldsinny population at Wrasse Reef is required.

References

Collins, K.J., Free, E.K., Jensen, A.C. and Thompson, S. (1993) Analysis of Poole Bay, U.K. lobster data. *International Council for the Exploration of the Sea, Mariculture Committee.* **1993/K**, 48.

Costello, M.J. (1991) Review of the biology of wrasse (Labridae: Pisces) in northern Europe. *Progress in Underwater Science* **16**: 29–51.

Darwall, W.R.T., Costello, M.J., Donnelly, R. and Lysaght, S. (1992) Implications of life history strategies for a new wrasse fishery. *Journal of Fish Biology* **41** (Supplement B): 111–23.

Hilldén, N.-O. (1981) Territoriality and reproductive behaviour in the goldsinny, *Ctenolabrus rupestris* L. *Behavioural Processes* **6**: 207–21.

Sayer, M.D.J., Gibson, R.N. and Atkinson, R.J.A. (1993) Distribution and density of populations of goldsinny wrasse (*Ctenolabrus rupestris*) on the west coast of Scotland. *Journal of Fish Biology* **41** (Supplement A): 157–67.

Sayer, M.D.J., Gibson, R.N. and Atkinson, R.J.A. (1995) Growth, diet and condition of goldsinny, *Ctenolabrus rupestris* (L.) on the west coast of Scotland. *Journal of Fish Biology* **46**: 317–40.

Seber, G.A.F. (1982) *The Estimation of Animal Abundance and Related Parameters* (2nd edn). Charles Griffin & Co. Ltd., London.

Chapter 6
Distribution and abundance of wrasse in an area of northern Norway

K. MARONI and P. ANDERSEN *Akva Instituttet AS, Avd. Flatanger, 7840 Lauvsnes, Norway*

Most published accounts of wrasse distribution in northern Europe, have given 64–65°N as the approximate northern limit. It was therefore thought unlikely that any quantities of wrasse would be found in Norway north of Trondheim. During the summer of 1992, a trial fishery was conducted in the county of Nord–Trøndelag (63.5–65°N). In total about 5000 goldsinny (*Ctenolabrus rupestris*) were caught, with a few cuckoo (*Labrus mixtus*). The number of fishes caught per unit effort was used as an indicator of population strength in different regions. As a result of this trial fishery, many salmon farmers in the region started to use wrasse for lice control in 1993 with very good results.

6.1   Introduction

Flatanger is a small municipality in the middle to northern part of Norway (64.5°N). Salmon production in the region last year was about 1500 tons. From 1990, sea lice (Copepoda, Caligidae) infestation was a major problem for the local salmon farming industry. To treat and control sea lice many methods were employed. On one occassion, nine Nuvan® treatments were necessary on the same smolt generation from May to October. Other methods of lice control have been onions and garlic (no significant effect) and freshwater treatment (the lice disappeared, but the smolts died from furunculosis). Lastly, during the autumn 1991 we were actively involved in the development of hydrogen peroxide (H_2O_2) as a treatment against lice, which proved more effective. Had hydrogen peroxide not worked, the whole fish farming activity in the municipality would have been eradicated during the autumn 1991 as a result of sea lice infestation.

In 1991 the use of wrasse as a lice treatment was considered. The only problem was that nobody had seen wrasse in this area of Norway before. Although local fishermen thought it unlikely that significant numbers were present, when a salmon farmer tried fishing for wrasse he caught some goldsinny [*Ctenolabrus rupestris* (L.)]. During spring 1992, a project was initiated to determine distributions and densities of wrasse populations in the county of Nord-Trøndelag.

6.2 Materials and methods

Two fishermen were employed to do most of the fishing. In addition, 16 fish farms in the county were involved in the project. The project was organized in close cooperation with the Norwegian Institute of Marine Research in Bergen (Å. Bjordal) and A/S MOWI, a Norwegian fish farming company (P.G. Kvenseth).

A total of 196 bow-nets were used: of these, 12 were normal Norwegian-produced eel-pots; 28 were Danish-produced eel-pots; and 156 were specially designed wrasse-pots (Bjordal, 1993). The fish farms each got about six pots, and the rest were used by the two wrasse fishermen. All fishing was recorded in a standard manner, covering map reference for each location, the kind of fishing gear used, depth of fishing, description of bottom conditions (sand, gravel, boulders, hard rock), scale of vegetation, topography, and type of bait used. Fishing dates were also recorded, as were the times that each fishing period started and finished.

The county was divided into separate zones, and strict rules were followed to reduce the risk of transferring disease from one area to another. Each fish farm only fished in their immediate vicinity.

6.3 Results

6.3.1 *Wrasse capture and length distribution*

In total, 5541 wrasse were caught during the project period lasting from the end of July to the end of September 1992. The active fishing period differed a little between regions (Table 6.1).

Almost 100% of the fish caught were goldsinny, with only a few cuckoo wrasse (*Labrus mixtus* L.) of both sexes. The numbers of wrasse caught, and the catch

Table 6.1 Fishing efficiency in different areas during the fishery mapping in 1992.

Area (Latitude)	Period (1992)	Pot-days[1]	Number of fish caught	Number of fish per pot per day
Nærøy/Vikna (64.8–65.0°N)	27.07–29.09	713	331	0.5
Fosnes (64.7°N)	12.08–30.09	627	498	0.8
Namsos (64.5°N)	07.08–30.09	186	24	0.1
Flatanger N (64.5°N)	18.08–05.10	351	1754	3.1
Flatanger M (64.5°N)	27.07–10.08	454	422	0.9
Kjelvika (64.4°N)	28.08–25.09	224	813	3.6
Flatanger S (64.4°N)	11.08–12.10	438	1723	3.9
Total		2993	5541	1.9

[1] Pot-days is defined as the number of days by pots in the sea, multiplied by the number of pots used each day.

per unit effort differed between regions, but the higher catch rates occurred at the more southern sites. In addition, almost the same number of fish were caught during the autumn after the project had finished.

The length/frequency distributions were skewed towards smaller fish in the northern part of the survey area.

6.3.2   *Vegetation*

Most wrasse were caught in areas with moderate to dense vegetation. In Nærøy/ Vikna, large areas have little vegetation and wrasse capture rates were low (Table 6.1).

6.3.3   *Topography*

Fish capture rates were higher where shores were steep compared to beach areas. Steep shores with a shelf at 2–5 m depth were recorded as good fishing areas.

6.3.4   *Type of fishing gear*

Of the three types used it was evident that the wrasse-pot was most effective. The two eel-pots were less suitable since cod and crab could also enter, and would attack and eat some wrasse.

6.3.5   *Bait and fishing time*

Crushed, fresh crab seemed to give best results. After two to three hours fishing time the effect of the bait seemed to diminish. Fishing time did not result in an increase in the number of wrasse caught. Some pots were inspected after two to three hours in the sea, and checked again after a subsequent period. Fishing efficiency for the wrasse pots diminished after two to three hours in the sea.

6.4   **Discussion**

Previously published accounts for goldsinny wrasse have given the northern limits of their distribution as ranging from about 61°N (Wheeler, 1978), to between 64.5 and 65°N (Wheeler, 1969; Quignard & Pras, 1986, respectively). This survey in a region ranging from 64 to 65°N has shown quite large goldsinny populations in this area of Norway. The lower catches per unit effort in the north of the region suggest that these populations may be close to the northern limit for goldsinny, although small numbers have been caught further north. The low catches of cuckoo wrasse again suggest that this region represents their northern limit. This accords with Wheeler (1969, 1978) who gave 64–65°N as the northern limit for cuckoo, but is south of the 68°N limit given by Quignard and Pras (1986).

Catches in the northern part of Nord-Trøndelag have increased from 1992 to 1994, possibly reflecting increased fishing intensity. To prevent over-exploitation of what may be more fragile populations compared with southern ones, routine tagging surveys and length-frequency distribution analyses are being undertaken on a regular basis in the region. As a result of this trial fishery, many salmon farmers in the region started to use wrasse for lice control in 1993 with very good results.

References

Bjordal, Å. (1993) Capture techniques for wrasse (Labridae). *International Council for the Exploration of the Sea: Fish Capture Committee* ICES-CM-1993/B, **22**: 3.

Quignard, J.-P. and Pras, A. (1986) Labridae. In *Fishes of the North-eastern Atlantic and the Mediterranean* Vol. II (Ed. by Whitehead, P.J.P., Bauchot, M.-L., Hureau, J.-C., Nielsen, J. and Tortonese, E.). UNESCO, Paris, pp. 919–42.

Wheeler, A. (1969) *The Fishes of the British Isles and North West Europe*. Macmillan, London.

Wheeler, A. (1978) *Key to the Fishes of Northern Europe*. Frederick Warne, London.

Chapter 7
Capture techniques for wrasse in inshore waters of west Scotland

J.W. TREASURER *Marine Harvest McConnell, Lochailort, Inverness-shire PH38 4LZ, UK*

The total catch of wrasse in Scotland in 1994 was estimated as 150 000 fish. The selectivity of prawn creels, traps and fyke nets for wrasse species and length composition were assessed. The species composition of catch in fyke nets in an Argyll sea loch in August 1994 was 41% goldsinny, 46% rock cook, 13% corkwing and 1% ballan, compared with 7%, 87%, 3% and 3% respectively on the Isle of Jura. While goldsinny dominated in experimental and commercial fisheries from Mull to Loch Ewe, rock cook comprised a greater proportion of catch south of Mull. Length frequency distribution in fyke nets in Argyll was in the range 80–144 mm for goldsinny and 80–154 mm for rock cook, but fish <110 mm were under-represented in prawn creels. The proportion of the wrasse catch <100 mm in length returned by fishermen to the water varied from 23–44%. The age range of goldsinny and rock cook captured in west Scotland was 3–16 years and 3–9 years respectively, with no evidence of dominant year classes. The instantaneous mortality rates for goldsinny and rock cook in Ardnamurchan in 1990–91 were 0.186 and 0.419 per annum respectively. Catch-per-unit effort in Argyll from 260 fyke nets set for one day was 0.52 per net for goldsinny, 1.16 for rock cook, 0.17 for corkwing, and 0.04 for ballan wrasse.

The catching power and selectivity of creels and traps are described and possible guidelines for the fishery proposed.

7.1 Introduction

Interest in wrasse as cleaner fish on Atlantic salmon, *Salmo salar* L., farms in Scotland commenced in 1989. Fishing for wrasse in Scotland commenced in 1990 and, initially, fish farmers tried to catch their own. However, the investment in suitable fishing gear was expensive and they did not have time, so contracts were awarded to fishermen to supply wrasse. The fishery is seasonal as few fish are captured at temperatures less than 7–8°C (Darwall *et al.*, 1992; Sayer *et al.*, 1993) and farms request wrasse for stocking from June to August when the smolts are small and when wrasse can still clean fish effectively before the onset of colder temperatures. Although the present study describes the fishery purely in terms of wrasse returns, there is an interesting by-catch associated with the fishing techniques employed. Saithe [*Pollachius virens* (L.)], pollock [*Pollachius pollachius* (L.)], cod (*Gadus morhua* L.), conger eel [*Conger conger* (L.)], scorpion fish [*Myoxocephalus scorpius* (L.), *Taurulus bubalis* (Euphrasen)], five-bearded rockling [*Ciliata mustela* (L.)], flatfish of various species, and dogfish [*Scyliorhinus canicula* (L.)] are frequently captured along with wrasse.

This account details an analysis of the wrasse fishery on the west coast of Scotland, and of capture techniques employed there. In making this analysis the following questions were addressed:

(1) Does the commercial fishery give an adequate representation of the distribution, species composition and relative abundance of wrasse in west Scotland?
(2) Is the fishing gear selective for species composition and length frequency distribution?
(3) What factors limit the fishery?

7.2 Materials and methods

7.2.1 *Study areas*

The frequency of each species and lengths of wrasse were measured in the catches of commercial fishermen, and in experimental fisheries as described in Treasurer (1994a). Fishing localities were grouped in three geographical zones: the north-west of Scotland, including the Isle of Skye and Lochs Torridon and Ewe, 120 km north of Oban; Ardnamurchan, including sea lochs from Loch Ailort to Loch Linnhe; and Argyll from Craignish Point to Jura (Fig. 7.1). In addition, detailed new data were collected from an Argyll sea loch and from the Isle of Jura (Fig. 7.2) in August 1994 to examine the influence of site topography and macroalgal cover on capture success. The south-east shore of the Argyll sea loch was exposed, with a gently sloping shore comprising rocks of boulder size, with cover of *Fucus* sp., and subject to wave action. On the north shore the shoreline was a rock-face of about 5 m dropping vertically into water, sheltered from wave exposure. There was a rapid drop in water depth to 3 m with a band of kelp about 5 m from the shore. At the entrance to the sea loch, the habitat was a gradually sloping rocky shoreline with kelp cover, both on the mainland and on small rocky outcrops. The habitat on the east coast of Jura was similar with exposure to wave action.

7.2.2 *Description of fishing gears*

Initially prawn creels (Fig. 7.3) were used to catch wrasse in Scotland. They were baited with shellfish, and fished close inshore in water of 3–10 m for 30–60 minutes on a rising tide. These were later used unbaited and fished overnight.

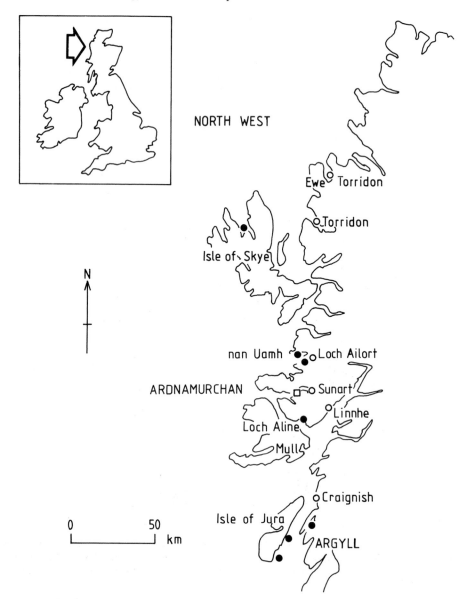

Fig. 7.1 Map showing the fishing/sampling locations for wrasse on the west coast of Scotland. Inset and arrow shows position in the British Isles. Key to fishing gear used: ○, 18 mm mesh prawn creels; ●, fyke nets; □, Sunart trap.

Creel dimensions were 57 × 40 × 30 cm. The entry aperture on the modified prawn creel was 40 mm and mesh size 12 mm bar compared with 18 mm on the original unmodified creel, although the large mesh size has mainly been used by fishermen. Creels are rarely used now but they are still utilized on farms for retrieving wrasse from cages.

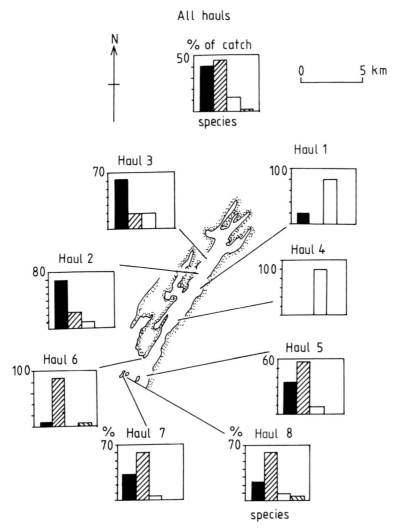

Fig. 7.2 Species composition of wrasse in fleets of 20 fyke nets in an Argyll sea loch. Number of fish examined (*n*) in each haul (H) were: H1:5, H2:57, H3:58, H4:7, H5:49, H6:39, H7:48, H8:33; hauls of all eight fleets of 20 fyke nets combined:296. Key: ■ goldsinny; ▨, rock cook; ▢, corkwing; ▧, ballan.

In Loch Sunart, a cylindrical polypropylene trap was developed, 60×30 cm with a 24 cm funnel tapering to a 40 mm 'eye' (Fig. 7.3). This has 12 mm rigid mesh and the fish are therefore less likely to escape as was sometimes observed for prawn creels. The Sunart traps are fished unbaited in fleets of 10, close inshore.

Fyke nets are now the main fishing gear in Scotland, and were first used for catching wrasse in Norway (Rae, *pers. comm.*). Typically, the nets used have leader netting with holding traps at each end. The leader is weighted on the

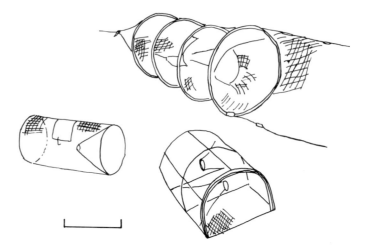

Fig. 7.3 Fishing gear used for wrasse in Scotland: top, fyke net with leader netting; bottom left, Sunart trap; bottom right, prawn creel. Scale bar = 30 cm.

bottom with buoyancy on the upper edge. The traps comprise up to seven plastic hoops of 50 cm diameter with a mesh size of 11 mm. The traps are fitted with guards to prevent otters entering the nets. Normally ten of these double fyke nets (counted as 20 fykes) are linked in fleets set parallel to the shore in water of 3–10 m depth, covering over 100 m length of fishing ground. These are normally close inshore on rocky ground with a cover of kelp or *Fucus* sp.

The species composition and total lengths (TL mm) of wrasse in commercial and experimental fisheries were measured. In the more detailed survey in the Argyll sea loch, fleets of 20 linked fyke nets were set for one day parallel to the shore in water of 3 m depth determined by echo-sounder. The number of each wrasse species per haul, lengths in selected hauls, and the proportion of undersize fish returned to the water were recorded. Catch per unit effort data were expressed as mean catch per net set day^{-1}.

7.3   Results and discussion

7.3.1   *Wrasse distribution and relative abundance*

Catches in commercial and experimental fisheries were dominated by goldsinny, *Ctenolabrus rupestris* (L.) north of the Isle of Mull (56°35′N) and rock cook, *Centrolabrus exoletus* (L.), south of the Sound of Mull (Treasurer, 1994a). For example, in Loch Ewe goldsinny formed 94% of the catch in prawn creels, and 93% in fyke nets in Loch Ailort. There were variations in this general pattern: for example, in Loch Ewe, only 2.4% of the catch was rock cook, while in nearby Loch Torridon rock cook comprised 36% of the catch. Data are presented elsewhere on species composition of wrasse in commercial fisheries in Scotland (Treasurer, 1994b). The breakdown of catches by species in the Argyll sea loch, as a percentage of total catch in one fleet of 20 fyke nets, set parallel to the shore, is shown in Fig. 7.2. The numbers of wrasse captured varied between localities. Catches on the exposed south east shore were low (hauls numbered 1 and 4) and consisted mainly of corkwing. Goldsinny was the dominant species on the sheltered north shore, comprising from 62–68% of catch in hauls 3 and 2 respectively. Rock cook was the main species at the more exposed locations at the entrance to the sea loch comprising from 57–87% of the catch. The overall catch composition in all hauls in the sea loch was: goldsinny 41%; rock cook 46%; and corkwing 12.5%, of a total of 296 wrasse captured.

On the east coast of Jura there were few goldsinny, the mean percentage composition of catch for 5 hauls being goldsinny 7.3%, rock cook 87%, corkwing 3.1% and ballan 2.6% from a total of 192 fish captured. Few ballan were captured in either the Argyll sea loch or Jura, and no cuckoo wrasse, *Labrus mixtus* L., were recorded.

7.3.2   *Catch composition using different fishing methods*

Species composition in catches using fyke nets and unmodified 18 mm mesh creels were compared at three localities. In Argyll, the two main species were represented in both gears (Fig. 7.4), although rock cook were more abundant in fyke nets compared with goldsinny, and corkwing comprised 12.5% of catch. Goldsinny dominated catches in Ardnamurchan, with 93% in fyke nets and 83% in prawn creels. More corkwing were present in creels, though low sample numbers make comparison difficult. In the north-west, goldsinny dominated fyke net catches, 72% compared to 56% in creels; and rock cook comprised 28% of fyke catches and 36% of the creel catch.

7.3.3   *Length frequency distributions using different gears*

The length frequency distribution of goldsinny wrasse in fyke net catches in the Argyll sea loch was 80–144 mm, similar to the range of 85–144 mm from fyke net catches in Skye (Fig. 7.5). In contrast, only length groups of 110 mm plus were retained in creels. The length range for goldsinny in the Sunart trap was 75–149 mm as the slightly rigid 12 mm mesh captured fish not retained by the other gears. The length frequency distribution of rock cook captured by fyke net in Argyll was 85 to 134 mm, similar to the range of 90–149 mm in Skye (Fig. 7.5). In the 18 mm mesh creel in Argyll, rock cook were 95–149 mm total length. Rock cook captured by the Sunart trap in Ardnamurchan, were 75–139 mm length; therefore smaller length groups were retained by this gear that were not represented in fyke nets or creels.

Insufficient corkwing were captured to present a length frequency distribution

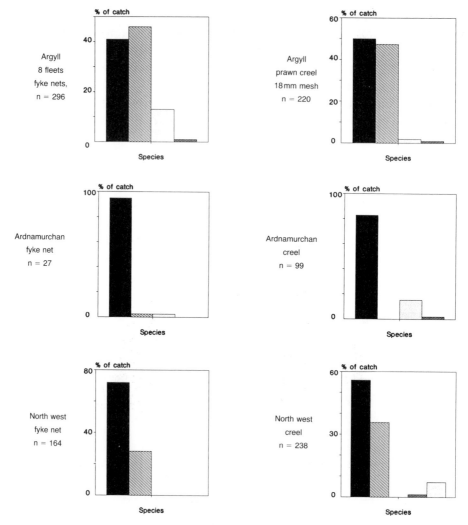

Fig. 7.4 A comparison of species composition in fyke net and prawn creel catches. Gear type and n are shown. Key: ■ goldsinny; ▨, rock cook; ☐, corkwing; ◩, ballan; ▨, cuckoo wrasse.

but, in the Argyll sea loch, corkwing were in the range 120–152 mm and ballan 135–192 mm.

The length frequency distributions for goldsinny in Scotland compare with 60–140 mm for goldsinny captured in 'shrimp pots' in west Ireland (this volume, Chapter 9).

7.3.4   *Gear selectivity*

The catching power of creel and Sunart trap depended on fish being attracted to and entering the trap and this may be affected by factors such as season, temperature (Darwall *et al.*, 1992), weather and reproductive condition. It has also been suggested that wrasse may be able to escape from creels (Chapter 9, this volume). In contrast, catches in fyke nets were mainly dependent on the leader net guiding the fish into a holding bag. Apart from the similarity in the species composition of catches in a given locality using the three methods, the length frequency distribution was also similar. The poor representation of smaller length groups in creels was explained by mesh selection.

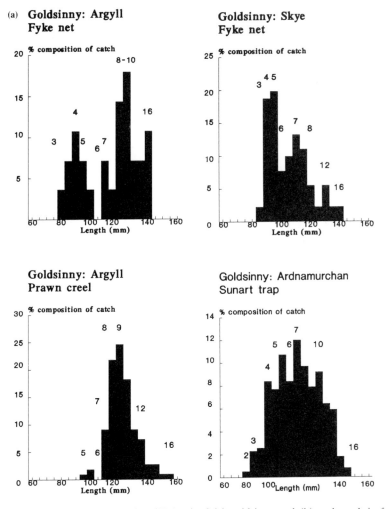

Fig. 7.5 Length frequency distribution (TL/mm) of (a) goldsinny and (b) rock cook in fyke nets, creels and traps. The estimated age of goldsinny from otoliths and rock cook from opercula are shown above the bars. Number of fish measured (*n*) = goldsinny: Argyll fyke 28, Argyll creel 110, Skye fyke 91, Sunart trap 391; rock cook: Argyll fyke 66, Argyll creel 104, Skye fyke 45, Sunart trap 83.

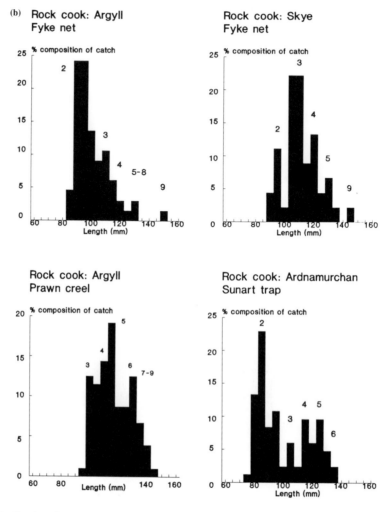

Fig. 7.5 *Continued*

7.3.5   *Reject rate of catch*

The length composition of wrasse is important: from the fish farmer's viewpoint there is little point in buying wrasse that will escape from salmon nets. The minimum length of goldsinny and rock cook retained by salmon net mesh of various sizes was examined and was found to be 90 mm for goldsinny in 12 mm mesh (Fig. 7.6). To give a safety margin in order to prevent fish squeezing through the mesh and becoming 'gilled', wrasse of a minimum length limit of 100 mm are accepted from fishermen by farmers in Scotland.

In two fleets of fyke nets set in the Argyll sea loch, 22.5% and 43.6% of wrasse, mainly goldsinny and rock cook, were judged by fishermen to be undersized and were released (Table 7.1); at Jura this was 31.5%. These measurements were

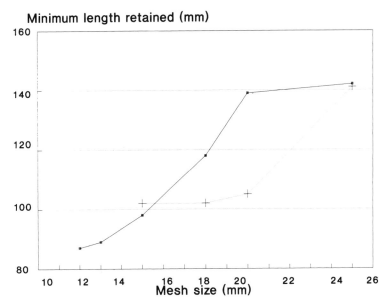

Fig. 7.6 Minimum length of goldsinny and rock cook wrasse retained by salmon net of six mesh sizes. No data available for rock cook wrasse smaller than 102 mm. Key: ■ goldsinny; +, rock cook.

Table 7.1 Proportion of wrasse (goldsinny and rock cook combined) captured in fyke nets and then released, because they were considered <100 mm total length. The accuracy of the size-grading determined by measuring lengths of retained and released fish is also shown.

Locality	Total length (mm)	Number retained	Number released	Total Number	% released
Argyll sea loch, haul 1		38	11	49	22.5
haul 2		20	17	37	43.6
Jura		61	28	89	31.5
Size grading					
Argyll sea loch	>100	48	4		
	<100	2	40		
Jura	>100	31	3		
	<100	4	25		

checked by measuring the retained and rejected catch (Table 7.1): 48 of 52 retained fish were >100 mm and four released fish were 100 mm, 102 mm, 103 mm and 100 mm respectively. Of fish <100 mm, 40 were released and two retained, both 97 mm. At Jura, three of 34 wrasse >100 mm were released and could have been retained and four of 29 fish <100 mm should have been released (97 mm, 95 mm, 95 mm and 96 mm).

The percentage of goldsinny >100 mm in two hauls in the Argyll sea loch (n = 28 fish measured) was 71.4%. The proportion was 99% (n = 110) (Fig. 7.5) using prawn creels with 18 mm mesh at Craignish (Treasurer, 1994b). The percentage of goldsinny catch >100 mm was 72.9% (n = 391) using Sunart traps in Ardnamurchan (Fig. 7.5) and 59.3% (n = 91) using fyke nets in Skye.

The proportion of rock cook of >100 mm length in catches with fyke nets in the Argyll sea loch was 48.5% (n = 66), compared with 99% (n = 104) using 18 mm prawn creels at Craignish and 34.9% (n = 83) using the Sunart trap (Fig. 7.5).

The proportion of wrasse catch returned by fishermen can be compared with length frequencies of goldsinny captured in baited shrimp pots in Lettercallow Bay, west Ireland (Chapter 9, this volume). Only 25.4% and 11% of goldsinny exceeded 100 mm in 1991 and 1992 respectively. In a study in Scotland, (this volume, Chapter 8), Sayer *et al.* found that 42.4% of mature males and 25.9% of mature females captured in baited prawn creels at Millport, Firth of Clyde were >100 mm. Goldsinny at Oban were faster growing and consequently 86.7% of males and 57.7% of females were removed by the fishery (this volume, Chapter 8).

As length frequency distributions vary between localities depending on fish growth, the effect of fishing on wrasse stocks will vary greatly, and any proposed conservation measures need to be determined in each situation by representative sampling of catches.

7.3.6   *Variability of catches*

There was great variation in the number of fish captured between individual nets set and between fleets of 20 nets. The total numbers of wrasse varied between net fleets from five to 89 fish, goldsinny from zero to 39, and rock cook zero to 78 (Table 7.2). The variance of catches was greater than the mean catch. The catches were log transformed as $y = \log(x + 1)$ before statistical analyses were carried out. For all wrasse, the arithmetic mean (AM) catch from 13 fleet settings was 37.4 and the geometric mean (GM) catch was 27.5 with 95% confidence limits (CL) from 16.7 to 44.8. For goldsinny, values were AM = 11.2, GM = 4.39, CL = 1.6–10.0; for rock cook, AM = 23.2, GM = 12.4, and CL = 5.5–26.2.

To examine changes in catch per unit effort many settings of net fleets would have to be used. For confidence limits to be below half or twice the geometric mean, seven fleets of 20 nets would be required in August (Fig. 7.7a). The catching power of nets could vary seasonally, as in pike, *Esox lucius* L., (Bagenal, 1972) and in perch, *Perca fluviatilis* L., and charr, *Salvelinus alpinus* (L.) (Craig & Fletcher, 1982), when using multimesh gill nets.

From 100 individual net settings in the Argyll sea loch, the arithmetic mean catch for all wrasse was 2.51 per net, 1.11 for goldsinny and 1.15 for rock cook. The geometric mean catch per net was 1.8, CL = 1.4–2.2 for all wrasse species combined, 0.69, CL 0.49–0.92 for goldsinny and 0.70, CL = 0.51–0.93 for rock

Table 7.2 Numbers (*n*) and % catch of wrasse of each species captured in fleets of 20 fykes in Argyll.

Fleet No.	Goldsinny		Rock cook		Corkwing		Ballan		Total *n*
	n	%	*n*	%	*n*	%	*n*	%	
1	1	20	0	0	4	80	0	0	5
2	39	68	13	23	5	9	0	0	57
3	36	62	11	19	11	19	0	0	58
4	0	0	0	0	7	100	0	0	7
5	17	35	28	57	4	8	0	0	49
6	3	8	34	87	0	0	2	5	39
7	16	33	29	60	3	6	0	0	48
8	8	24	20	61	3	9	2	6	33
9	2	3	62	95	1	2	0	0	65
10	2	20	7	70	0	0	1	10	10
11	0	0	6	43	4	29	4	29	14
12	10	11	78	88	1	1	0	0	89
13	0	0	14	100	0	0	0	0	14
Total *n*									
Sea loch fleets 1–8	120	41	135	46	37	13	4	1.4	296
Jura fleets 9–13	14	7	167	87	6	3	5	3	192

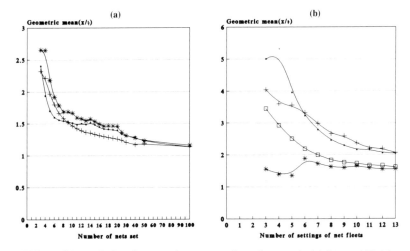

Fig. 7.7 95% confidence limits of geometric mean catches of wrasse in (a) fleets of 20 fyke nets and (b) individual fyke nets. The mean is divided or multiplied by the value given to estimate lower and upper confidence limits. Key: ■, goldsinny; ▢, rock cook; ★, all species including small numbers of corkwing and ballan.

cook. The number of net settings using individual nets for confidence limits to be below half or twice the geometric mean would be at least five (Fig. 7.7b).

7.3.7   Catch per unit effort (CPUE)

Comparative data are difficult to obtain as baited prawn creels were fished from 30–60 minutes to one day and as unbaited creels were also used. The data for the Sunart trap were compared with fyke nets in Argyll (Table 7.3) but fishing effort for these was also not equivalent as traps were fished over three days, compared with one day using fyke nets. Catches in pots, for example, fyke nets and traps for flatfish (Van der Veer *et al.*, 1992), and in shrimp pots for wrasse (Chapter 9, this volume) have been found to be inversely related to period fished, with catches declining markedly with time fished.

   These observations should be borne in mind when comparing the catch data for fyke nets and the Sunart trap in west Scotland, and also with other locations (Table 7.3). The CPUE figures from Lettercallow Bay, in the west of Ireland, were 0.25 goldsinny and 0.04 corkwing per baited shrimp pot-hour fished (this volume, Chapter 9), 0.11–0.51 goldsinny per baited prawn creel hour in west Scotland (this volume, Chapter 8), and 0.1–3.9 goldsinny $pot^{-1} day^{-1}$ in Norway this volume, Chapter 6).

7.3.8   Age distribution

In examining the fishery there is a need to relate age to the length/frequency distributions of fish to determine the age range of fish captured. In a previous study (Treasurer, 1994a), scales, opercula bones, and sagittal otoliths of the five wrasse species found in west Scotland were examined to determine the best means of ageing these fish. Goldsinny were most accurately aged from sagittal otoliths and the other species from opercula. It is not possible to make generalizations about length at age (Fig. 7.8) as growth varies with location and with the sex of fish (see also this volume, Chapter 8), and females may mature earlier than males (Darwall *et al.*, 1992). This earlier investment in gonadal tissue in females may occur at the expense of somatic growth, and may explain why males are longer at any specific age than females. For example, female goldsinny in Ardnamurchan were 83 mm length at age three years compared with 93 mm in males. In some localities no male or female goldsinny attained the retention length of 100 mm until they were at least five years old (Fig. 7.8a). Growth of rock cook also differed between localities and males were also consistently longer at a given age. These differences in growth between localities and sexes indicate that possible effects of exploitation on wrasse stocks could vary between areas.

   Using age determinations of a random sample of goldsinny and rock cook captured in the Ardnamurchan area in 1990 and 1991, an age distribution of fish in catches was constructed (Fig. 7.9). Goldsinny populations were dominated by

Table 7.3 Catch per fyke-day in an Argyll sea loch (160 nets) and Jura (100 nets), compared with the Sunart trap in Ardnamurchan (60 nets).

Locality	Fishing Gear	Goldsinny	Rock cook	Corkwing	Ballan	Total No.	Reference
Argyll:							
sea loch	fyke net	0.75	0.84	0.23	0.03	1.85	This study
island	fyke net	0.14	1.67	0.06	0.05	1.92	This study
Ardnamurchan	Sunart trap	0.89	0.08	0.007	0	0.98	This study
Norway	fyke nets and wrasse traps	$0.1–3.9\ \text{day}^{-1}$					This volume, Chapter 6
Western Ireland	baited pots	$0.25\ \text{h}^{-1}$					This volume, Chapter 9
Western Scotland	baited prawn creels	$0.11–0.51\ \text{h}^{-1}$					This volume, Chapter 8

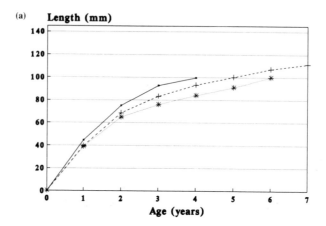

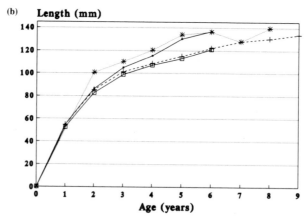

Fig. 7.8 Mean length-at-age of (a) goldsinny and (b) rock cook wrasse back-calculated from sagittal otoliths and opercula respectively. Key: (a) ■, Ardnamurchan males; −+−, Ardnamurchan females; ··*··, males and females combined, north west area. (b) ■, Ardnamurchan males; +, Ardnamurchan females; *, Ardnamurchan males and females combined, note these are observed not back-calculated values; □, males and females combined means, back-calculated values. Observed lengths at age include additional growth beyond the last annulus.

younger age classes but longevity was to 16 years and all age classes were represented. This can be compared with an age range of one to twelve years in west Ireland (Chapter 9, this volume). Rock cook is a more short lived species, up to nine years-old. Three- and four-year-olds dominated the population in west Scotland but all age-groups were represented (Fig. 7.9). The instantaneous mortality rate (Z) was calculated by plotting $\log_e$ percentage composition of catch on age. In goldsinny, Z was 0.186 per annum and in rock cook it was 0.419. In this calculation, the smaller age classes prone to escaping through the net meshes were excluded, that is less than four-years-old in goldsinny and less than three-years-old in rock cook. This can be compared with Z values of 0.425 in 1991 and

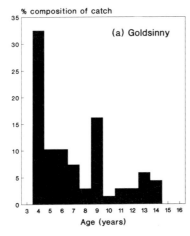
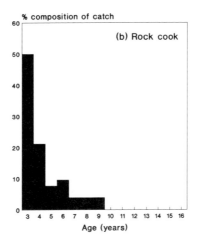

Fig. 7.9 Age structure of (a) goldsinny (68 fish) and (b) rock cook wrasse (52 fish) captured in prawn creels in Ardnamurchan in August and September 1990 and 1991. Goldsinny <4 years and rock cook <3 years were not included as these were not adequately retained by the 12 mm mesh.

0.672 in 1992 in fished goldsinny stocks in Lettercallow Bay, western Ireland (Chapter 9, this volume).

The ratio of males to females was 1.22 in goldsinny of less than three-years-old, not significantly different from unity ($\chi^2 = 0.80$, $P > 0.05$, $n = 80$) and 0.83 in rock cook, also not significant ($\chi^2 = 0.38$, $P > 0.05$, $n = 42$). Sex ratios in goldsinny were similar to 1 : 1.1 and 1 : 1.12 for this species captured at Oban, west Scotland in 1992 and 1994 respectively (this volume, Chapter 8).

7.3.9   *Guidelines for the fishery*

The size of the fishery in 1994 was estimated as 150 000 wrasse from a question-naire circulated to fish farmers (Chapter 14, this volume). No specific regulations govern the fishery, but possible guidelines are proposed here. These reflect not only influencing factors highlighted in the present study but also conditions of contract applied by major fish farming companies in Scotland:

(1) The catching power of creels, traps and fyke nets is unknown and varies seasonally (Darwall *et al.*, 1992).

(2) The fishery in Scotland is seasonal, mainly from June to August, as wrasse are stocked when smolts are put to sea and can only be captured in any number when temperatures are greater than 7°C (Darwall *et al.*, 1992; this volume, Chapter 8).

(3) Wrasse should not be captured in the immediate vicinity of other salmon farms, to reduce the possibility of transmitting disease.

(4) Also, for this reason, wrasse should preferably be captured in the same sea loch/locality as the farm where they are to be stocked.

(5) No ballan should be used as these can be aggressive to salmon, and the relatively rare cuckoo wrasse should also be returned.

(6) Health screening of a representative sample of wrasse is advised prior to stocking.

(7) Fish of less than 100 mm length should be returned as they are not retained by 12 mm salmon net mesh.

(8) No extensive exploitation of a given area should be undertaken, and continuous moving of fishing gear is recommended.

Acknowledgements

I am grateful to Steve MacDonald and Ian McWhinney for assistance and also commercial fishermen, including Rodney George, Douglas Wilson and Douglas Lamont.

References

Bagenal, T.B. (1972) The variability of the catch from gill nets set for pike *Esox lucius* L. *Freshwater Biology* **2**: 77–82.

Craig, J.F. and Fletcher, J.M. (1982) The variability in the catches of charr, *Salvelinus alpinus* L., and perch, *Perca fluviatilis* L., from multimesh gill nets. *Journal of Fish Biology* **20**: 517–26.

Darwall, W.R.T., Costello, M.J., Donnelly, R. and Lysaght, S. (1992) Implications of life-history strategies for a new wrasse fishery. *Journal of Fish Biology* **41** (Supplement B): 111–23.

Sayer, M.D.J., Gibson, R.N. and Atkinson, R.J.A. (1993) Distribution and density of populations of goldsinny wrasse (*Ctenolabrus rupestris*) on the west coast of Scotland. *Journal of Fish Biology* **43**: (Supplement A), 157–67.

Treasurer, J.W. (1994a) The distribution, age and growth of wrasse (Labridae) in inshore waters of west Scotland. *Journal of Fish Biology* **44**: 905–18.

Treasurer, J.W. (1994b) Distribution and species and length composition of wrasse (Labridae) in inshore waters of west Scotland. *Glasgow Naturalist* **22**: 409–17.

Van der Veer, H.W., Witte, J.I., Beumkes, H.A., Dapper, R., Jongejan, W.P. and Van der Meer, J. (1992) Intertidal fish traps as a tool to study long-term trends in juvenile flatfish populations. *Netherlands Journal of Sea Research* **29**: 119–26.

Chapter 8
The biology of inshore goldsinny populations: can they sustain commercial exploitation?

M.D.J. SAYER[1], R.N. GIBSON[1] and R.J.A. ATKINSON[2] [1] *Centre for Coastal and Marine Sciences, Dunstaffnage Marine Laboratory, P.O. BOX 3, Oban, Argyll PA34 4AD; UK and [2] University Marine Biological Station, Millport, Isle of Cumbrae KA28 0EG, UK*

The effectiveness of wrasse as cleaner-fish has resulted in their commercial exploitation on the west coast of Scotland over the past four to five years. If the goldsinny fishery was to require regulation in the future, quantitative information on many aspects of its biology would be essential. Present knowledge of the distribution and density of goldsinny populations on the west coast of Scotland is reviewed and discussed here. To assess the possible effects of exploitation, an experimental fishery for goldsinny wrasse (*Ctenolabrus rupestris*) was established and regularly worked for three seasons by creeling with modified prawn creels. The status of the fishery was estimated by comparing catch per unit effort between successive years. Effects on the population structure caused by exploitation and possible sources of replacement were assessed by comparing the size distributions and sex ratios of goldsinny caught during the first season with those caught during the third. To act as a control, goldsinny were captured at a nearby second site, measured and returned every year. The results of this study are considered in relation to the goldsinny fishery. The effect on goldsinny populations of limiting the fishery to fish of 100 mm total length or larger is also discussed.

8.1   Introduction

The goldsinny [*Ctenolabrus rupestris* (L.)], is one of the north European wrasse species (Labridae) that are effective in the control of sea lice (Copepoda: Caligidae) infestation of farmed Atlantic salmon (*Salmo salar* L.) (Bjordal, 1991; Costello, 1993; Treasurer, 1993). Goldsinny attracted most of the initial attention with respect to lice control and, because it is the most common wrasse in many areas of the west coast of Scotland, is subject to substantial commercial exploitation. Prior to their use in salmon farms, goldsinny were considered to be of little economic value. Consequently, the effects of exploitation, both on the fish and its habitat, are not understood and, as yet, cannot be predicted. If the fishery was to require regulation, then a good understanding of many aspects of the biology of goldsinny would be necessary. This need was addressed by a research programme funded jointly by the Scottish Salmon Growers Association and the Crown Estate. This account reviews some of the research relating to goldsinny population densities and dynamics undertaken during that programme. It also presents data from experimental fisheries that were established in order to estimate replacement rates, to examine the sustainability of the fishery, and to assess the possible ecological effects of exploitation.

8.2   Goldsinny population densities and dynamics

Four sites were surveyed in detail for nearly two years (May 1992 to March 1994) to assess seasonal changes in the density and structure of goldsinny populations (Table 8.1). Each site was clearly defined, either on the basis of its physical appearance (e.g. a long horizontal crevice) or by use of fixed rope transects. Diving surveys were conducted at each site in an identical fashion, covering the same route and in the same direction on each occasion. To eliminate any differences in counting technique, the same diver was responsible for all the surveys. Surveys were repeated at different times of the day and states of tide. As well as total numbers of goldsinny, it was also possible to record the numbers of young-of-the-year. Sexual identification by underwater observation was not always possible outside the reproductive period (June–August). Surface water temperatures were taken regularly at both Dunstaffnage and Millport.

Numbers of goldsinny observed at each of the four main study areas are illustrated in Fig. 8.1. In both years, the seasonal trends for abundance were similar for all four sites. Numbers observed increased from April/May, were high

Table 8.1 Physical details of the four main study sites where records of goldsinny population density and structure were made.

Location of study	Habitat type	Depth of water	Area
Isle of Cumbrae (Millport)			
A6	Vertical rock cliffs, with areas of rock/boulder scree (0.2–1.5 m diameter). Macroalgal cover to 3 m.	8–10 m	102.8 m^2
A4	Boulder scree at 45° (0.1–0.7 m dia.). Macroalgal cover to 3 m. Horizontal sand bottom at 7–8 m.	5–7 m	48.0 m^2
Eilean Mhôr (Dunstaffnage)			
C3	Horizontal boulder field (0.1–1.2 m dia.) at base of vertical rock cliffs. Macroalgal cover to 6 m.	16–20 m	51.8 m^2
C5	Deep horizontal crack in vertical cliff. Macroalgal cover to 6 m.	10–13 m	35.7 m*

* Total length of crevice entrance. Area was difficult to estimate as only a small fraction of the available habitat was visible.

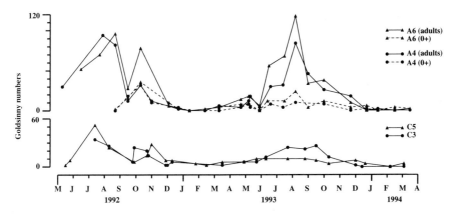

Fig. 8.1 Goldsinny numbers on each transect at four sites, two on the Isle of Cumbrae (A6 and A4) and two at Eilean Mhôr (C5 and C3) (May 1992 to March 1994). Broken lines represent numbers of 0+ goldsinny.

during the summer months, and declined to near negligible levels over the winter months. The highest densities recorded were at site A4 (3.9 goldsinny m^{-2}); sites A6 and C3 had maximum densities of 1.3 and 1.1 fish m^{-2} respectively. Area at site C5 was difficult to measure. During July 1992, 1.5 goldsinny m^{-1} of crack entrance were recorded. As the height of the crack ranged from 0.09–0.32 m, considerable concentrations of wrasse were present at the entrance at that time of year (*c.* 5–17 fish m^{-2}). Newly-settled goldsinny (0+) were only observed at the two Cumbrae sites (A4 and A6). In 1992, they were first observed in late August, and by September accounted for over 50% of the goldsinny recorded at those sites. In 1993, newly-settled goldsinny were not observed until the end of September, almost a month later than in 1992. No 0+ animals were recorded at sites C3 and C5 even though considerable reproductive activity had been observed at both sites during the early summer. Subsequent surveys away from C3 and C5 found the nearest 0+ goldsinny over 300 m from either site in shallow water (3–6 m), within macroalgal cover, on a boulder scree habitat.

The main factor limiting distribution of goldsinny was the availability of suitable refuge in the rocky sublittoral. The preferred refuge type consisted of holes between or under rocks with more than one entrance/exit (as previously noted by Costello, 1991). The greatest densities of goldsinny were observed in habitats of this type. Substantial numbers were also recorded in caves and crevices. The goldsinny was by far the most common fish at all the sites visited in this study, and was always present where some suitable refuge existed. Goldsinny were, however, largely absent at depths likely to be affected by freshwater run-off. Where they were present at shallower depths, the refuge was highly complex and possibly afforded some protection against rapid and marked changes in temperature and salinity. Although considered generally to be a purely marine fish, and absent from estuarine waters (Costello, 1991), Quignard (1966) found that goldsinny fed at salinities as low as 12‰ at temperatures of 18–20°C. However, temperature reduces the salinity tolerance of fish (Finstad *et al.*, 1988; Malloy & Targett, 1991) and it is possible that the salinity tolerance of goldsinny reported by Quignard (1966) is less at lower water temperatures more applicable to the west coast of Scotland. Exposure of goldsinny to a series of environmentally-relevant temperature and salinity regimes suggests that goldsinny–but not corkwing [*Crenilabrus melops* (L.)], or rock cook [*Centrolabrus exoletus* (L.)]–have a high tolerance of low salinity and low temperature (minima 8‰ and 4°C respectively) (this volume, Chapter 10). It is therefore likely that the distribution pattern for goldsinny reported by Sayer *et al.* (1993) is more a factor of optimum refuge availability than the direct influence of winter water conditions.

Young-of-the-year goldsinny were most commonly observed at depths shallower than 18 m, and were typically found on rocky boulder scree in a depth range of 2–7 m. Quignard and Pras (1986) also reported that adult goldsinny were found deeper than juveniles; a similar effect was observed in other temperate reef fish populations (Leum & Choat, 1980; Bell, 1983; Buxton & Smale, 1989). Macroalgal cover did not determine the presence of 0+ goldsinny. Some 0+ fish were observed in or around macroalgae, but greatest densities were recorded deeper than its lower limit. Shallow areas of boulder scree may act as nursery areas for goldsinny, because 0+ and possibly 1+ fish were not observed at deeper sites. Juveniles probably migrate away from these nursery areas after their first winter so replacement to the adult population comes from this migration from nursery areas rather than from postlarval settlement in the adult habitat. If juvenile survival is density-dependent then exploitation might result in increased juvenile survival.

The numbers of goldsinny observed actively swimming were highly seasonal with high numbers recorded from June to October in 1992 and 1993 and near zero from December to May (Fig. 8.1). Seasonal change in the activity of goldsinny (Costello, 1991; Treasurer, 1991; Darwall *et al.*, 1992), and other temperate rocky reef fishes (Kotrschal, 1983) is well known. Although Hilldén (1981) explained low winter numbers by offshore migration, Costello (1991) and Darwall *et al.* (1992) suggested that during the winter goldsinny remain inactive and hidden within refuges. The use of anaesthetics to flush fish from amongst rocks during the winter has confirmed the latter suggestion because numbers of captures/ observations per unit area in winter were similar to summer values (Sayer *et al.*, 1994). In areas where the available refuge was less complex, individual goldsinny were observed in a motionless state, wedged into shallow rock crevices. Similar behaviour must occur at boulder sites during the winter but it is difficult to observe.

Marked increases and decreases in numbers of active goldsinny visible to the diver were recorded in May and November respectively. Darwall *et al.* (1992) suggested that goldsinny are only active in Irish waters when seawater temperatures exceed 7°C. Sayer *et al.* 1993 estimated goldsinny disappearance to occur at water temperatures of 8.7°C and reappearance at 8.2°C. From measurements of metabolic activity over a temperature range of 4–10°C, goldsinny only become hypometabolic below 6°C (this volume, Chapter 10). It is unlikely therefore that observed winter disappearance at 7 or 8°C is caused by inactivity. Goldsinny may remain active at or above temperatures of 6°C because they feed at times of the year when temperatures are below the disappearance/reappearance range (Sayer *et al.*, 1995). Disappearance at or below 8°C is probably a result of goldsinny changing their behaviour from swimming in open water to the occupation of cryptic refuges.

8.3 Experimental fishery for goldsinny

Underwater surveys by sub-aqua diving were used to determine the limits of distribution of isolated goldsinny populations on Maiden Island (Oban) and in Dunstaffnage Channel. At Maiden Island, marks were painted on the cliff face to denote the population distribution limits. The island site was fished regularly using creels baited with crushed sea urchin from April to December 1992. Limited fishing was also carried out from April to September 1993. Fishing resumed in June 1994 through to August of that year. At the channel site, baited creels were set by diver on an isolated boulder scree patch. Catches at both sites were quantified as number of goldsinny caught per creel hour. Total length and sex of all fish caught were recorded. Captured goldsinny from Maiden Island were removed from the area and used in other studies. After measuring length and/or weight, sexing, and tagging (TBF-2 tags, Hallprint, Australia) goldsinny from the channel site, the fish were re-released onto the exact area of capture by diver.

At the exploited site, catches per unit effort (CPUE) were high during July to September 1992 (Table 8.2). A total of 125 goldsinny were taken from the site during this period in 1992. In 1993, 54 goldsinny were captured from July to September, but at a significantly reduced CPUE compared with the respective 1992 months ($P < 0.05$, Student's *t*-test). The CPUE July to September 1994 was significantly lower than catches for the same periods in 1992 or 1993 ($P < 0.01$ and 0.05 respectively). The male:female ratio in 1994 (1:2.00) was not significantly different from than that of 1992 (1:1.18; χ^2, $P > 0.05$). The mean total length (mm) of both males and females was not significantly different ($P > 0.05$; Student's *t*-test after tests for homogeneity of variance) for both 1992 and 1994 catches. Captured goldsinny were not sexed in 1993.

At the site where fished animals were returned after sexing and measuring, 33 goldsinny were captured in 1993 with a similar fishing effort in 1992 when 32 goldsinny were caught. The male:female sex ratio of 1:2.33 in 1993 was not significantly different from that of 1992 (1:1.46; χ^2, $P > 0.05$). The mean total lengths (mm) of the fish caught in the two years were not significantly different ($P > 0.05$).

There are potentially three sources of replacement of exploited goldsinny

Table 8.2   Catch per unit effort, sex ratio and mean length of goldsinny captured from exploited and unexploited fisheries from July 1992 to September 1994.

Period of catch	Exploited fishery (removal)	Unexploited fishery (tag/return)
	Catch per Unit Effort (goldsinny per creel hour; mean $\pm$ SE or single fishing effort)	
July–September 1992	0.51 $\pm$ 0.08	0.18
July–September 1993	0.27 $\pm$ 0.02	0.18
July–September 1994	0.11 $\pm$ 0.04	
	Sex Ratio (male : female)	
July–September 1992	1:1.18	1:1.46
July–September 1993		1:2.33
July–September 1994	1:2.00	
	Total length (mm; mean $\pm$ SE)	
July–September 1992	108.3 $\pm$ 1.6	120.2 $\pm$ 3.6
July–September 1993		121.6 $\pm$ 2.5
July–September 1994	110.8 $\pm$ 3.3	

populations: postlarval settlement, juvenile (0+/1+) migration from nursery areas, and adult migration from other populations. Juvenile goldsinny (0+/1+) were not observed or captured away from shallow water sites (Sayer *et al.*, 1993). Postlarval goldsinny may therefore settle preferentially in shallow water. Where juveniles remain at these nursery sites they may recruit to the local population, although age frequency analysis of goldsinny 1+/2+/3+ age-groups at these sites showed lower than expected 1+/2+ goldsinny numbers (Sayer *et al.*, 1995a). Two-year-old (plus) goldsinny were the youngest captured at the exploited site, and so migration of 1+ animals away from the site of postlarval settlement is therefore a possibility. Adult goldsinny migration has been recorded to virgin artificial reefs over distances greater than 3 km (this volume, Chapter 3), and may have occurred during the tagging experiments in this study, where adult returns were lower than expected, and at the exploited site where average size was maintained. Quantitative information relating to sources and rates of replacement is still lacking for the goldsinny fishery.

8.4   Can goldsinny populations sustain commercial exploitation?

The results from the experimental fishery demonstrated a decline in goldsinny CPUE with continued exploitation. However, those results were from a single site which may not be typical of the overall fishery, and were obtained over only three years.

There are a number of problems associated with attempts to recommend sustainable yields for goldsinny. Goldsinny examined from three areas of the west coast of Scotland have entirely different biologies (for example, in age/growth, diet, and GSI/HSI trends; Sayer *et al.*, 1995). Estimation of growth, fecundity, and mortality, which are essential factors for assessing fishery ecology, is therefore not practical on a general area basis. Local variation would have to be taken into account for all fisheries.

The following series of observations on goldsinny life history have been made during recent investigations of its biology (Costello, 1991; Darwall *et al.*, 1992; Sayer *et al.*, 1993):

(1)   Reproductive activity takes place at both deep and shallow water sites.
(2)   Goldsinny eggs and larvae are planktonic.
(3)   Postlarval goldsinny settle in shallow water areas.
(4)   Young-of-the-year goldsinny overwinter at these shallow water nursery sites.
(5)   One-year-old (plus) and some adult goldsinny may move away from, and/or between, certain areas or habitats.

Because shallow water areas appear to sustain higher adult densities (Sayer *et al.*, 1993), these areas must be considered to be the primary source of larval input. For a fishery, a conflict of interests therefore arises because most goldsinny are taken from shallow waters of 3–5 m (Treasurer, 1991). A possible suggestion is to limit fishing to depths of 15–25 m. A more practical one would be to limit mesh sizes on nets, or (as there is no evidence of mortalities on release of wrasse captured in shallow water) for commercial suppliers or customers to impose a minimum limit for wrasse length (Costello *et al.*, 1994). Costello *et al.* (1994) suggest a minimum limit of 10 cm total body length for wrasse introduced into salmon farm cages.

Geographic variation in age/length relationships for goldsinny results in estimates of age for a length of 10 cm, ranging from 2+ to 3+ years for males, and 3+ to 6+ years for females (Sayer *et al.*, 1995). Goldsinny become sexually mature at 2+ (Darwall *et al.*, 1992; Sayer *et al.*, 1995) and so a minimum length of 10 cm will ensure that a certain proportion of the available broodstock remains each year. This proportion will vary with location. Using the catch data of Sayer *et al.* (1995), a limit of 10 cm total body length resulted in 42.4% of mature males, and 25.9% of mature females being removed from Millport populations by exploitation. Because growth rates are higher at Oban, 86.7% of mature males and 57.7% of mature females were removed. The reasons for slower growth are unclear. Goldsinny with slower growth rates were captured by Sayer *et al.* (1995) from areas of higher population density and had markedly higher GSI values. If the magnitude of gonadal development affects growth rate and is dependent on population density (which in turn may be habitat-dependent), then exploitation, by reducing wrasse numbers, may cause a reduction of gonadal development, a concomitant increase in growth rate, and result in a larger proportion of the broodstock being captured.

It has been suggested that goldsinny populations would be resilient to exploitation due to their small size, maturation at an early age and production of large numbers of eggs (Darwall *et al.*, 1992). These criteria satisfy in part those suggested by Garrod and Horwood (1984) for rapid population recovery from exploitation. One other criterion proposed by Garrod and Horwood (1984) was rapid growth. If, as appears possible, goldsinny growth is linked to population density, habitat, gonadal development and exploitation, then it will not be easily possible to predict the effects of a fishery on population status. At present, the goldsinny fishery is, to a certain extent, self-regulating. The 10 cm limit imposed by the consumer should ensure that breeding populations will withstand continued exploitation, although with resultant alterations both to the population age/size structure and sex ratio. The present scale of wrasse exploitation, coupled with the complexity of the fishery and its environment, suggest that a degree of caution is required. If the use of wrasse by the salmon farming industry continues on a long-term basis, it would seem prudent to continue investigations into ways of producing wrasse in hatchery breeding programmes, in order to safeguard wild stocks.

Acknowledgements

This work was funded by Argyll and Islands Enterprise, the Scottish Salmon Growers Association and the Crown Estate.

References

Bell, J.D. (1983) Effects of depth and marine reserve fishing restrictions on the structure of a rocky reef assemblage in the north-western Mediterranean Sea. *Journal of Applied Ecology* **20**: 357–69.

Buxton, C.D. and Smale, M.J. (1989) Abundance and distribution patterns of three temperate marine reef fish (Teleostei: Sparidae) in exploited and unexploited areas off the southern Cape coast. *Journal of Applied Ecology* **26**: 441–51.

Bjordal, Å. (1991) Wrasse as cleaner-fish for farmed salmon. *Progress in Underwater Science* **16**: 17–28.

Costello, M.J. (1991) Review of the biology of wrasse (Labridae: Pisces) in northern Europe. *Progress in Underwater Science* **16**: 29–51.

Costello, M.J. (1993) Review of methods to control sea lice (Caligidae: Crustacea) infestations on salmon (*Salmo salar*) farms. In *Pathogens of Wild and Farmed Fish Sea Lice* (Ed. by Boxshall, G.A. and Defaye, D.). Ellis Horwood, Chichester, pp. 219–52.

Costello, M.J., Treasurer, J.W. and Sayer, M.D.J. (1994) A guide to the use of cleaner-fish. *Fish Farmer* **17**(6): 7–8.

Darwall, W.R.T., Costello, M.J., Donnelly, R. and Lysaght, S. (1992) Implications of life-history strategies for a new wrasse fishery. *Journal of Fish Biology* **41** (Supplement B): 111–23.

Finstad, B., Staunes, M. and Reite, O.B. (1988) Effect of low temperature on sea-water tolerance in rainbow trout, *Salmo gairdneri*. *Aquaculture* **72**: 319–28.

Garrod, D.J. and Horwood, J.W. (1984) Reproductive stategies and response to exploitation. In *Fish Reproduction: Strategies and Tactics* (Ed. by Potts, G.W. and Wootton, R.J.). Academic Press, London, pp. 367–84.

Hilldén, N.-O. (1981) Territoriality and reproductive behaviour in the goldsinny *Ctenolabrus rupestris* L. *Behavioural Processes* **6**: 207–21.

Kotrschal, K. (1983) Northern Adriatic rocky reef fishes at low winter temperatures. *Pubblicazioni della Stazione Zoologica di Napoli I: Marine Ecology* **4**: 275–86.

Leum, L.L. and Choat, J.H. (1980) Density and distribution patterns in a temperate marine fish *Cheilodactylus spectabilis* (Cheilodactylidae) in a reef environment. *Marine Biology* **57**: 327–37.

Malloy, K.D. and Targett, T.E. (1991) Feeding, growth and survival of juvenile summer flounder *Paralichthys dentatus*: experimental analysis of the effects of temperature and salinity. *Marine Ecology Progress Series* **72**: 213–23.

Quignard, J.-P. (1966) Recherches sur les Labridae (Poissons, Teleostéens, Perciformes) des côtes européennes: systématique et biologie. *Naturalia Monspeliensia (Zoologie)* **5**: 7–248.

Quignard, J.-P. and Pras, A. (1986) Labridae. In *Fishes of the North-eastern Atlantic and the Mediterranean* Vol. II (Ed. by Whitehead, P.J.P., Bauchot, M.-L., Hureau, J.-C., Nielson, J. and Tortonese, E.), UNESCO, Paris, pp. 919–42.

Sayer, M.D.J., Cameron, K.S. and Wilkinson, G. (1994) Fish species found in the rocky sublittoral during winter months as revealed by the underwater application of the anaesthetic quinaldine. *Journal of Fish Biology* **44**: 351–53.

Sayer, M.D.J., Gibson, R.N. and Atkinson, R.J.A. (1993) Distribution and density of populations of goldsinny wrasse (*Ctenolabrus rupestris*) on the west coast of Scotland. *Journal of Fish Biology* **43** (Supplement A): 157–67.

Sayer, M.D.J., Gibson, R.N. and Atkinson, R.J.A. (1995) Growth, diet and condition of goldsinny on the west coast of Scotland. *Journal of Fish Biology* **46**: 317–40.

Treasurer, J. W. (1991) Wrasse need due care and attention. *Fish Farmer* **14(4)**: 24–6.

Treasurer, J.W. (1993) Management of sea lice (Caligidae) with wrasse (Labridae) on Atlantic salmon (*Salmo salar* L.) farms. In *Pathogens of Wild and Farmed Fish* (Ed. by Boxshall, G.A. and Defaye, D.). Ellis Horwood, Chichester, pp. 335–45.

Chapter 9

The effect of intensive fishing of wild wrasse populations in Lettercallow Bay, Connemara, Ireland: implications for the future management of the fishery

S.J.A. VARIAN, S. DEADY and J.M. FIVES *Department of Zoology, Martin Ryan Institute of Marine Science, University College, Galway, Ireland*

The use of wrasse (Teleostei, Labridae) as cleaner-fish on salmon farms has led to the development of a new fishery for a number of species. A preliminary investigation into the impact of this fishery on two species of wrasse, goldsinny, *Ctenolabrus rupestris* (L.), and corkwing, *Crenilabrus melops* (L.), was carried out over two years at a small bay in the west of Ireland. Changes in population size were assessed from catch returns and the efficiency of the trapping method (shrimp pots) was analysed using catch per unit effort (CPUE) data. A significant ($P < 0.05$) decrease in mean CPUE was observed for both species in the second year of fishing, despite increased numbers of goldsinny captured as a result of increased fishing effort. The subsequent investigation of the impact of fishing on the population structure of the goldsinny, the main species captured, showed that there was a decline in the proportions of larger goldsinny from 1991 to 1992 and from April to July 1992. Catch curves showed that goldsinny greater than six-years-old were being rapidly depleted over the two years, suggesting that the older year classes were the most vulnerable to the fishery. It was found that young male goldsinny were the most important component of the fishery, because of their fast growth and high abundance. These results are discussed in relation to recovery and sustainability of the wild wrasse populations as well as the future management of the fishery.

9.1   Introduction

The use of wrasse (Teleostei, Labridae) as cleaner-fish on salmon farms has recently expanded from the original experiments in Norway (Bjordal, 1988, 1990; Costello & Bjordal, 1990) to commercially-sized salmon farms in Norway, Scotland and Ireland (Bjordal, 1992; Darwall *et al.*, 1992a; Treasurer, 1994a; Deady *et al.*, 1995). In these areas, five species of wrasse commonly occur, and of these, corkwing [*Crenilabrus melops* (L.)], rock cook [*Centrolabrus exoletus* (L.)] and

100

goldsinny [*Ctenolabrus rupestris* (L.)] have been found to be the most effective cleaners. Originally considered as a commercially unimportant group of fish (Muus & Dahlstrom, 1974), there is now a growing demand for live wrasse by salmon farmers. Treasurer and Henderson (1992) reported that over 100000 wrasse were captured in Scotland in 1991 and in Norway, 180000 wrasse were used on salmon farms in 1992 (Bjordal, 1993). The increasing requirement for more cleaner-fish to stock salmon farms has led to the development of a commercial fishery for wrasse (Darwall *et al.*, 1992b). As wrasse inhabit rocky inshore areas, fishing is limited to trapping methods such as fyke nets, prawn creels and shrimp pots. In Norway, a new folding wrasse trap has been developed to meet demands for sufficient numbers of live wrasse (Bjordal, 1993). To date, there is still a lack of information on this newly developed fishery, its impact on wild wrasse populations and factors affecting the efficiency of the trapping methods.

This paper presents results obtained from an intensive wrasse fishing survey in a previously unfished small bay in the west of Ireland. Corkwing and goldsinny wrasse were captured by means of shrimp pots during the summer months of 1991 and 1992 in order to supply a nearby salmon farm. This provided an opportunity to examine the population structure of goldsinny, the main species captured, and to assess the efficiency of the fishing method employed, as well as the impact of fishing on the population size of both species.

9.2   Study area, materials and methods

Lettercallow is a small, relatively sheltered bay, located on the south coast of Connemara in western Ireland (Fig. 9.1). The bay has a surface area of approxi-

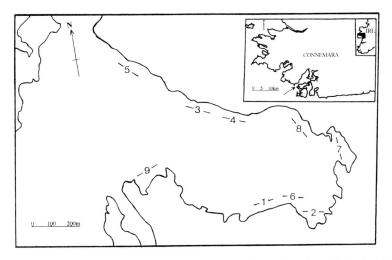

Fig. 9.1 Map of Lettercallow Bay, Connemara, Ireland, showing the locations of the nine sites fished for wrasse.

mately $8000\,m^2$ and a maximum depth of 14 m. The substrate varies from rocks and boulders, which dominate the shoreline and extend subtidally to approximately 10 m, to a sandy silty bottom which extends over the deeper areas and the centre of the bay.

Fishing was carried out using cylindrical shrimp pots which were 56 cm long and 38 cm in diameter, with an aperture at each end of 45 mm. The sides were covered with 10 mm mesh but the maximum bar mesh size at either end of the pots was 12 mm. The pots were assembled in maximum sets of ten, where each set consisted of a weight and a marker buoy, attached at either end of a single headrope along which the pots were connected at 8 to 10 m intervals by 5 m lengths of rope. They were baited with crushed mussel (*Mytilus edulis* L.) approximately every three days, were laid in water at a depth between 5 and 8 m and hauled after a minimum period of two hours. In 1991, five sites around the bay were fished (from July to October inclusive) and in 1992, a total of nine sites were fished (from March to July inclusive), including five sites from the previous year (Fig. 9.1). Records were kept of the site number, time at which the pots were laid and hauled and the species and number of wrasse collected. The wrasse were measured and then assigned to various 2 cm length-groups accordingly (e.g., 6–8 cm, 8–10 cm, etc.). Any fish smaller than 8 cm were returned to the site at which they were captured, whilst the remainder of the catch was placed into a holding cage (Polar circle, 10 m dia., 10 m depth, mesh size 8 mm bar) prior to stocking with the salmon. Daily temperature readings were taken throughout both fishing seasons.

The CPUE was calculated from the catch returns as the number of wrasse captured per pot-hour multiplied by 100 (where pot-hour was calculated as the total number of pots multiplied by the total number of hours fished). Only fishing intervals up to 24 hours duration were included in the analysis as recommended by Van der Veer *et al.* (1992).

In order to estimate the structure of the goldsinny population, a random sample of fish ($n = 132$) was used to determine the distribution of year classes and sexes throughout the length-groups (Ricker, 1958). Goldsinny were aged by means of their sagittal otoliths, which were soaked in 70% alcohol for approximately one hour before examination on a black background under a binocular microscope. The total length of the fish was measured to the nearest millimetre and its sex was determined from microscopical examination of the gonads. Changes in population structure over the two years were assessed using the length frequency data obtained from catch returns. However, this method only provided an estimate of how the dynamics of the wild goldsinny populations were influenced by fishing. In addition, the adult sex ratio was approximated, defined as the ratio of reproductively mature males to females. The impact of the fishing on the corkwing population structure was not assessed due to the low recorded abundance and erratic distribution of this species around the bay.

9.2   **Results**

In total, 4521 goldsinny were trapped in Lettercallow Bay between July 1991 and July 1992. Of these, 2465 fish of less than 8 cm in length (54.5% of the total catch) were returned to the wild stock, and 2056 (45.5% of the total catch) were retained for use as cleaner-fish. High losses of wrasse less than 10 cm were recorded in the commercial trials associated with the fishery, indicating that the wrasse were escaping through the nets (minimum mesh size 12 mm bar). The fishery was therefore selectively removing fish of greater than 10 cm length, with some redistribution of the 8–10 cm length-group around the bay. Only a small proportion of the goldsinny captured, 15.2% (686 fish) was greater than 10 cm total length.

Corkwing were not as abundant as goldsinny, with a total number of 361 fish caught over the two years. However, corkwing tended to be larger than goldsinny, with 92.0% (332 fish) over 10 cm and only 5.7% less than 8 cm. The total catch of corkwing in 1992 was lower than that of the first year despite an increased fishing effort (Table 9.1), while the numbers of goldsinny trapped in 1992 were higher than the previous year.

9.3.1   *Factors affecting CPUE*

CPUE declined in 1991 and 1992 as the length of fishing time (i.e. number of hours pots were laid down between hauls) increased (Fig. 9.2). CPUE was significantly and inversely correlated with the length of fishing time (Spearman's Rank Correlation Coefficient, $P < 0.001$). There was a significantly lower CPUE in the fishing periods greater than 10 hours (Mann-Whitney U test, $P < 0.05$) but an increase in the proportion of larger fish (Fig. 9.3).

The CPUE of goldsinny was significantly correlated with sea temperature both years (Spearman's Rank Correlation Coefficient, $P < 0.001$). In 1991, the mean

Table 9.1 Details of the wrasse fishing operations carried out in Lettercallow Bay in 1991 and 1992.

Fishing details	1991	1992
Total number of hauls*	1300	2316
Months fished	July–October	March–July
Sites fished	1–5	1–9
Total goldsinny catch	1313	3208
Total corkwing catch	186	175
Mean CPUE goldsinny (fish pot-hour^{-1} × 100)	25.6	18.1
Mean CPUE corkwing (fish pot-hour^{-1} × 100)	4.1	1.2

* 1 haul = 1 pot laid and pulled once.

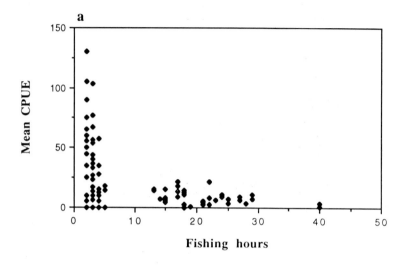

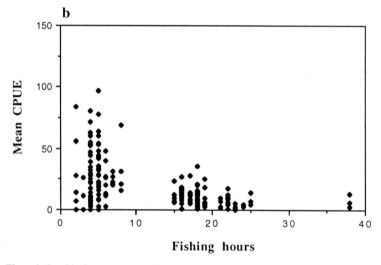

Fig. 9.2 The relationship between mean CPUE of goldsinny and length of fishing time in (a) 1991 and (b) 1992.

CPUE of the goldsinny dropped from a maximum value of 31.5 in July to a minimum of 5.2 in October, corresponding to the decline in temperatures (Fig. 9.4). In 1992, an increase in mean CPUE from March to July corresponded to the rise in temperatures.

Mean CPUE was greatest in the early morning and evening. However, this was not a significant factor in the fishery (Kruskal-Wallis test; $P > 0.05$). The relationship between CPUE and tidal movements was not significant (Mann-Whitney U test, $P > 0.05$). Greatest mean CPUE was recorded during low tide in 1991, while in 1992, the mean CPUE was greatest during periods of high tide.

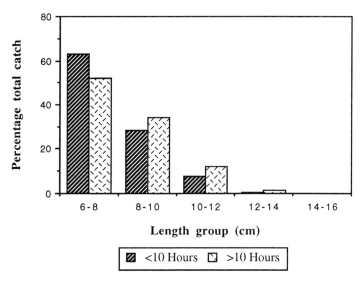

Fig. 9.3 The percentage frequency of goldsinny in each length group captured in periods less than or greater than 10 hours.

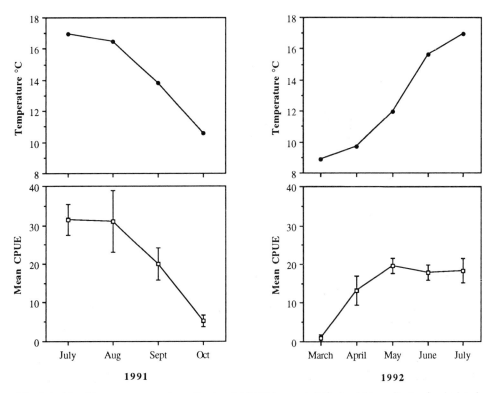

Fig. 9.4 Monthly mean temperatures (top) and CPUE (mean ± SE) of goldsinny (bottom) calculated for 1991 (left) and 1992 (right).

9.3.2   *Effect of fishing on population size and structure*

Abundance

Five of the nine sites were fished both years. Overall the mean CPUE of goldsinny and corkwing at these five sites decreased from 25.6 and 4.1 fish pot-hour^{-1} in 1991 to 18.1 and 1.2 fish pot-hour^{-1} in 1992 respectively (Table 9.1). The fall in CPUE was significant for both species (Mann-Whitney U test, $P <$ 0.005) despite an increase in fishing effort and numbers of goldsinny caught in 1992.

When relating fishing effort (number of hauls) over the two years to mean CPUE for each site (Fig. 9.5), it was evident that some sites were able to support a higher fishing intensity than others. For example, the greatest decline in the mean CPUE of goldsinny occurred at site 5 despite the fact that this site was fished less intensively than other sites.

Size frequency distribution

Because of the selectivity of the fishing gear, no fish less than 6 cm in length were recorded in the catches in Lettercallow. There was a decrease in the percentage of large goldsinny (>10 cm) caught over the two years (Fig. 9.6). The proportion of fish over 10 cm in length declined from 25.4% in 1991 to 11.0% in 1992. There was also a decline in the percentage frequency of the larger individuals observed between April and July in 1992 (Fig. 9.7a), with an associated decrease in CPUE (Fig. 9.7b). In July, no goldsinny over 12 cm were recorded in the catches.

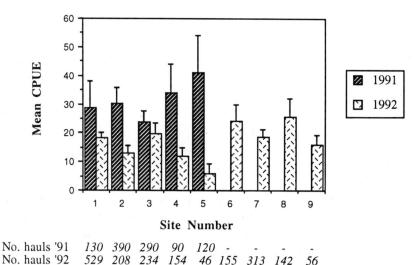

No. hauls '91   *130 390 290   90 120   -   -   -   -*
No. hauls '92   *529 208 234 154   46 155 313 142   56*

Fig. 9.5 Mean CPUE (+SE) of goldsinny calculated for each site fished in 1991 and 1992.

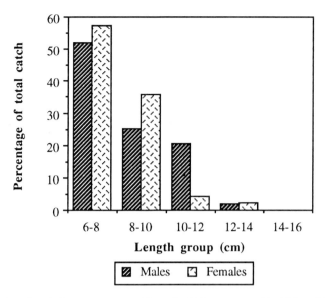

Fig. 9.6 Percentage length frequency composition of goldsinny captured in 1991 and 1992.

It was estimated that 78.6% of the total catch of large goldsinny (>10 cm) were male (Fig. 9.8). Males tended to be larger than females, as an estimated 22.7% of male goldsinny captured measured over 10 cm in length, compared with only 6.8% of females. There was an estimated decline in the percentage numbers of males caught over the two years, from 54.7% in 1991 to 51.6% in 1992, with a subsequent alteration of the sex ratio from 1.68:1 (male:female) in 1991 to 1.54:1 in 1992.

Age frequency distribution

Male goldsinny grew faster than females (Fig. 9.9a). 56.5% of males reached a length of at least 10 cm by the age of four, compared with only 2.6% of females. The bulk of the catch greater than 10 cm was formed by three to six-year-old fish, comprising 86.0% of the large goldsinny (>10 cm) caught in Lettercallow (Fig. 9.9b), and 83.9% of these were estimated to be male. However, the third year class was of greatest value to the fishery, as it constituted 39.2% of the exploitable catch, with approximately 95.0% male.

The age frequencies exhibited by the total catches of goldsinny for 1991 and 1992 are presented in Fig. 9.10. More young fish (one to six-year-olds) were captured in the second year of fishing than in the first, but fewer older fish (seven to twelve-year-olds). There was also an increase in instantaneous mortality (Z) from 0.425 in 1991 to 0.672 in 1992.

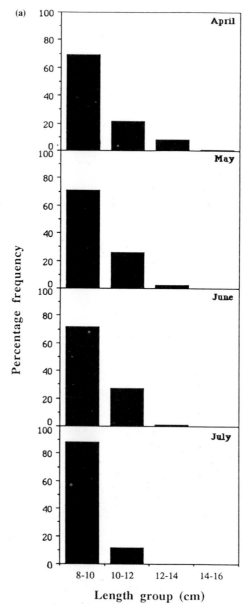

Fig. 9.7   (a) Percentage frequency distribution of goldsinny length groups for each month in 1992; (b) mean (±SE) CPUE of goldsinny for each length group (>10 cm total length) for each month in 1992.

9.4   Discussion

The cleaner-fish industry had a minimum size requirement of 10 cm (total length), since smaller wrasse escaped through the mesh of the salmon nets (Deady *et al.*, 1995). For the same reason, Darwall *et al.* (1992b) also recommended that only

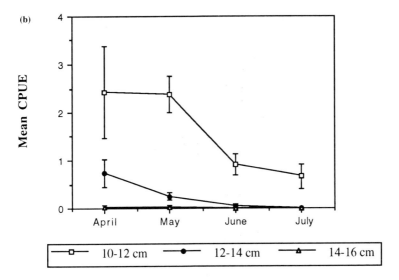

Fig. 9.7 *Continued*

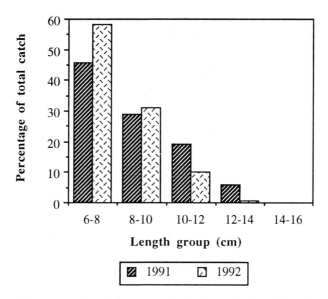

Fig. 9.8 Estimated percentage length frequency distribution of male and female goldsinny for the combined 1991 and 1992 catch.

wrasse greater than 10 cm length were suitable for stocking with salmon. In this study the use of shrimp pots proved to be an effective method of wrasse capture, providing sufficient numbers of wrasse ($n = 1018$) to supply the salmon cages, which were stocked with approximately 60 000 salmon.

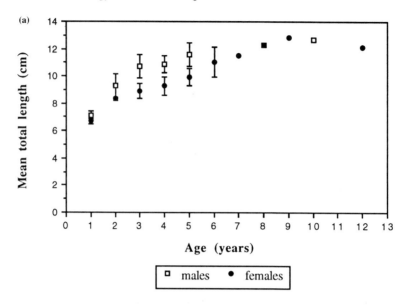

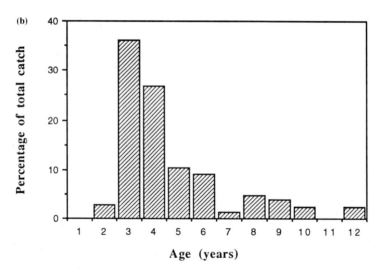

Fig. 9.9 (a) Mean (±SD) length at age of sexed goldsinny; (b) percentage age frequency distribution for the total catch of large goldsinny (>10 cm).

Goldsinny was by far the most abundant wrasse species in the bay (*n* = 4521), with a maximum density of 2 goldsinny m^{-2} observed by SCUBA divers (Darwall & Lysaght, *pers. comm.*). However, a comparatively small proportion of the goldsinny catch in Lettercallow was exploitable by the fishery (15.2% ≥ 10 cm). Other studies which have recorded the length frequency distributions in catches from Scotland and Norway (Treasurer, 1994b, Chapters 6, 7, this volume) reported at least 50% of goldsinny greater than 10 cm length. This difference may be

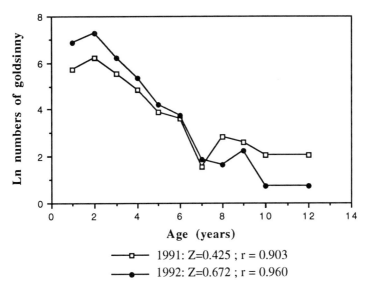

Fig. 9.10   Changes in the age frequency distribution of the goldsinny catch as calculated for 1991 and 1992.

attributed to the nature of the sites fished rather than variation in the type of fishing gear used. Pots used in this study had a bar mesh size of 10–12 mm, similar to the fishing gear employed in Scotland and Norway, where fyke nets and traps had a mesh size of 10 mm and 12 mm respectively (Treasurer, 1994b; this volume, Chapters 6 and 7). Sayer *et al.* (1993) found areas ranging in depth from 6–10 m to be preferred sites for goldsinny nurseries. Lettercallow has a maximum depth of 13 m, and though the shrimp pots selected fish greater than 6 cm, some one-year-old goldsinny were recorded in the bay, suggesting that it may be a nursery ground for juvenile wrasse. Presumably, as a result, there would be a high recruitment of young goldsinny to the bay, accounting for the high proportion of small fish recorded in the catch. In addition, in a survey of Lough Hyne in Ireland, divers noted that smaller (<10 cm length) goldsinny tended to be distributed at shallower depths (<10 m) than larger fish (Lysaght & Darwall, *pers. comm.*). The fishing conducted in Scotland and Norway was conducted in depths of up to 20 m (Treasurer, 1994b; Chapters 6, 7, this volume). Therefore, one would expect a higher proportion of large fish inhabiting these areas than in Lettercallow.

   Corkwing were more suited to the fishery in terms of size, with 92% of the catch measuring over 10 cm. However, it was erratically distributed around the bay with low recorded abundance. Additional corkwing could not be obtained since fishing was confined to areas within the immediate vicinity of the farm in order to minimize possible transfer of disease, a precaution currently taken by farms using cleaner-fish in Ireland (Darwall *et al.*, 1992b).

The effectiveness of the fishing method employed was reflected in the CPUE, where there was an observed decrease in CPUE with increasing fishing time. A decline in fishing efficiency with increasing time has also been reported for other fish trapping methods (Van der Veer *et al.*, 1992). The high CPUE recorded in the first few hours of fishing indicates that wrasse enter the pots soon after placement in the water. The subsequent decline in CPUE with increased fishing time, together with the increase in the proportion of larger fish in catches exceeding 10 hours, suggests that the smaller fish may leave the pots if left undisturbed for longer periods. Small goldsinny have been observed to reach satiation level more rapidly than larger goldsinny when fed with mussel in aquaria (Varian, *pers. obs.*). It is therefore possible that smaller goldsinny in the shrimp pots feed faster and, since goldsinny are crevice-dwelling fish (Darwall *et al.*, 1992b; Turner & Warman, 1991), leave through one of the apertures without difficulty, before the larger fish. However, they may also be forced out by the aggressive behaviour of larger territorial males. Territoriality of goldsinny in the wild has been described by Hilldén (1981) and intra-species aggression, including mouthfighting, has been observed in the wild (Hilldén, 1981), in salmon cages and in aquaria (Varian, *unpubl. data*).

In 1992, fishing was carried out during the spawning season. Hilldén (1981) observed reduced foraging behaviour due to courtship activity when goldsinny were reproductively active. If the catching power of the pots was affected by the lower feeding intensity of the goldsinny during this period, an associated decrease in the numbers of reproductively active fish captured would have been expected. Male and female goldsinny were found to be sexually mature at 2+ years in Lettercallow (Varian, *unpubl. data*), where mean total lengths were 9.32 cm and 8.30 cm respectively. Although there was an observed reduction in goldsinny greater than 10 cm in the second year of fishing, the 8–10 cm length group showed a percentage increase in the length frequency composition of the catch. Therefore, courtship activity was probably not a significant factor affecting the fishery.

The ability of some sites to support a higher fishing intensity than others demonstrates that the distribution and densities of wrasse around the bay were not uniform. The main factor limiting distribution of goldsinny has been found to be the availability of suitable refuges in the rocky sublittoral (Sayer *et al.*, 1993; Darwall & Lysaght, *pers. comm.*). Therefore, any variation in the arrangement of the rocky substrate would affect the distribution and density of the fish along the shoreline of Lettercallow. Another factor which could have limited their distri-bution is the presence of freshwater run off, since Sayer *et al.* (1993) found goldsinny to be largely absent at depths likely to be affected by it. However, there were no observations of such sites made in Lettercallow.

Groups of five or six individuals, particularly goldsinny, were frequently taken in a single pot, while other pots in the same set remained empty. Goldsinny and corkwing forage within territories, which are usually less than $2\,m^2$ and $10\,m^2$ respectively (Sjolander *et al.*, 1972; Hilldén, 1981; Potts, 1985). A number of wrasse may reside in one territory since both wrasse species exhibit female-defence polygyny, where one territorial male may guard a harem of up to five females (Hilldén, 1981; Dipper, 1981; Potts, 1985). It is therefore likely that the proximity of a pot to a territory could govern its catch efficiency.

Darwall *et al.* (1992b) observed reduced numbers of wrasse in the wild at temperatures below 7°C. Until recently it was believed that wrasse migrated to deeper water in the winter (Fiedler, 1964; Hilldén, 1981). However, it has now been shown that wrasse undergo a period of 'hibernation' at low temperatures since they have been observed by divers in a motionless state, wedged into rock crevices in winter (Sayer *et al.*, 1993, 1994; Darwall & Lysaght, *pers. comm.*; Deady & Varian, *pers. obs.*). Furthermore, in Scotland, the use of the anaesthetic quinaldine by divers to flush fish from amongst rocks during winter, showed that the numbers/observations per unit area in winter were similar to summer values (Sayer *et al.*, 1993, 1994). Growth and feeding intensity have also been found to be related to temperature (Quignard, 1966; Dipper *et al.*, 1977; Deady & Fives, 1995), reflecting reduction in wrasse activity at low temperatures. Therefore the wrasse fishing season is limited to the warmer months of the year, when temperatures are over 7°C. This approximately corresponds to the study period between March and October, in Lettercallow.

Goldsinny demonstrate strong diel behaviour and are largely inactive during the hours of darkness (Lysaght, *pers. comm.*; Turner & Warman, 1991). The higher values of CPUE found in the early morning and evening in Lettercallow probably correspond to the high concentration of hauls taken at these times as opposed to any trend in wrasse feeding activity. Turner and Warman (1991) observed peaks in activity of goldsinny at dawn and dusk in the wild, whilst an increased and more variable abundance was observed by Darwall and Lysaght (Darwall & Lysaght, *pers. comm.*). However, in the cleaner fish trials at Lettercallow, both goldsinny and corkwing were observed to clean and forage most actively at dawn and midday in the salmon cages, with a significant decrease in activity at dusk (Deady *et al.*, 1995).

Catch per unit effort provides an outline of fish abundance and comprises the main abundance index for many fisheries (Richards & Schnute, 1992). It can therefore be assumed that any decline in CPUE in this study would reflect a corresponding decrease in the size of the wrasse population in the bay. The significant decrease in CPUE of both species from 1991 to 1992 and the increase in instantaneous mortality over the two years indicates that exploitation rates in the bay were possibly exceeding replacement rates. Reduced abundances of wild wrasse stocks were also observed by divers in 1992 (Darwall & Lysaght, *pers. comm.*). There may have been some improved recruitment of 6–8 cm goldsinny to the fishery with the onset of growth following winter in 1992. However, the reduction in the relative proportion of larger wrasse in the catch (i.e. fish greater than 10 cm in length), is a likely symptom of removal by the fishery of large fish from the wild. The pattern of reduced CPUE and lower percentage frequency of larger wrasse in the catch returns is described as a characteristic trend of over-fishing by Munoz (1991). It was also evident within the five-month fishing period in 1992, demonstrating that over exploitation of wrasse stocks within a confined area, such as Lettercallow Bay, could occur in a relatively short time-span.

All the length groups removed by the fishing gear were well represented in the random sample of goldsinny that were aged and sexed. Because the fishing gear selected fish greater than 6 cm length, there were no 0+ year old fish recorded. It is also probable that a large mean-length at age one year-old was due to selective loss of smaller one-year-old fish. Therefore, the abundance of juvenile fish within the bay could not be measured and natural recruitment and mortality rates of the population were not calculated.

The age frequency curves for the catches of goldsinny in 1991 and 1992 illustrate the rapid depletion of the older fish, their size and low natural abundance making them particularly vulnerable to overexploitation. However, it is evident that the bulk of the exploitable catch was composed of younger fish (three to six-year-olds, an estimated 86.0%), the majority of which were male (an estimated 83.9%). The slower growth and resulting smaller size of the females made them less susceptible to exploitation than the males, with calculations estimating only 21.4% of the total goldsinny catch greater than 10 cm to be female. Other studies determining age and growth of goldsinny have also reported faster growth in males than in females (Quignard, 1966; Hilldén, 1978; Treasurer, 1994c; Sayer *et al.*, 1995). There also seems to be some geographical variation in wrasse growth rates (Quignard, 1966; Hilldén, 1978; Treasurer, 1994c; Sayer *et al.*, 1995), which would suggest some subsequent differentiation in potential recovery of exploited populations.

Estimated adult sex ratios for the total catches of goldsinny in 1991 and 1992 were male biased with the male:female ratio at 1.68:1 in 1991, and 1.54:1 in 1992 (presuming that the fishing gear selected both males and females with equal intensity). Skewed adult sex ratios in a polygonous breeding system may be caused and maintained by differential mortality of the sexes due to predation, or habitats with different suitabilities for each sex, combined with a high degree of philopatry (Kodricbrown, 1988). Male reproductive success depends both on the number of males competing for mating opportunities, and the number of repro-ductively active females. Consequently, skew in the adult sex ratios affects the intensity of sexual selection through changes in the variance in male reproductive success. Other studies have documented skewed adult sex ratios for species with polygynous breeding systems (Kodricbrown, 1988).

Since goldsinny and corkwing are highly territorial fish (Hilldén, 1981), the effect of male-male competition might result in the larger, competitively superior males maintaining territories and dominating the females, with the smaller, subordinate males becoming either sneaker or satellite males, adopting the alter-native non-territorial low-cost reproductive tactics described by Dipper (1981),

Hilldén (1981), Kodricbrown (1988), Darwall *et al.* (1992b) and Turner (1993). In his study of the sex ratio manipulation of male pupfish *Cyprinodon pecosensis* Echelle & Echelle, Kodricbrown (1988) found territorial males to acheive higher spawning success than satellite males, thus adopting a higher fitness breeding tactic. The removal of large male goldsinny from Lettercallow Bay by fishing might have been responsible for the change in adult sex ratios over the two years, from 1.68:1 to 1.54:1 (male:female). If so, the resulting reduced male-male competition might have allowed some of the small subordinate males to assume the position of the previously dominant males removed from the hierarchy, thus acheiving a higher spawning success. Had intensive fishing persisted in Lettercallow for an extended period (>2 years), there may have been continued reduction in the relative proportion of males in the catch, resulting in further manipulation of the sex ratio. Chan and Ribbink (1990) have shown that sneakers of the cichlid *Pseudocrenilabrus philander* can become territorial males (Turner, 1993), and Kodricbrown (1988) has observed satellite male pupfish acquiring territories in a femalebiased society. In this study, there would also have been an associated reduction in female choice, causing more females to spawn with males that might not be as genetically fit as the previous more competitive and successful males. The effects of this phenomenon on the state of the future goldsinny population are as yet unknown.

As mentioned above, the goldsinny matures at the age of two years (Varian, *unpubl. data*; Sayer *et al.*, 1995). It is a batch spawner and has planktonic eggs and larvae (Dipper, 1976; Darwall *et al.*, 1992b). Females in Lettercallow were found to produce an average of 12808 eggs per gramme of gonad weight per batch, where total length ranged from 8.3 cm to 12.8 cm (Varian, *unpubl. data*). The number of batches a female produces in one spawning season as well as the survival rate of the eggs and larvae are not known. The corkwing is a continuous spawner and has sticky demersal eggs that are protected by the male in a nest. The females mature at the age of 2+, whilst the males mature at the later age of 3+ (Dipper, 1976; Dipper & Pullin, 1979; Deady & Fives, *unpubl. data*). Corkwing larvae are also planktonic (Dipper, 1976; Dipper & Pullin, 1979). The reproduction of the two wrasse species is therefore characterized by early maturation and high fecundity, suggesting an ability to respond rapidly to population depletion (Darwall *et al.*, 1992b).

The removal of fish over 10 cm length diminishes the stock of mature spawning adults. However, since 90.9% of females aged were aged 5+ before they reached a length of 10 cm, the majority would probably have spawned in the wild for at least three years before they were removed by the fishery. Because most of the males would have been exploited by the fishery at the younger age of two or three years, they would probably have had only one or two spawning seasons prior to capture. However, the larger size of the corkwing may make it more vulnerable to the fishery than the goldsinny since the females may reach the minimum size requirement of the fishery more quickly. Also, the removal of nest-guarding male corkwing may have a detrimental effect on the survival of their eggs (Darwall *et al.*, 1992b).

The larvae of both species have been recorded in Galway Bay, including areas which are in close proximity to Lettercallow Bay (e.g. Kilkieran Bay and Golam Head; between two and six miles away respectively) (Fives, 1967, 1970; Dunne, 1970, 1972; Fives & O'Brien, 1976; Cheetham, 1981; Tully, 1986; Doyle, 1988). It has also been sugggested that goldsinny eggs may be amongst the most abundant fish eggs in the plankton in Irish waters (Kennedy & Fitzmaurice, 1969; Kennedy *et al.*, 1973). Therefore the immigration of planktonic stages into Lettercallow Bay is possible. However, Doyle (1988) recorded the abundance of goldsinny and corkwing larvae from the plankton in Galway Bay, during 1972 to 1977 and goldsinny was found to be sporadically abundant whilst corkwing was observed to range from low to intermediate levels. These results suggest that larval recruitment of goldsinny and corkwing to the fishery from outside the bay could be variable. The fact that Lettercallow appears to be a nursery ground for juvenile wrasse suggests that there would be larval recruitment to the population within the bay as well as from outside the bay. The territorial nature of both species, particularly goldsinny, which can remain in the same territory for several years (Hilldén, 1981), may affect the degree to which immigration of older fish into the bay would replenish stocks.

The early maturity and high fecundity of the two wrasse species suggests that populations would be resilient to fishing (Darwall *et al.*, 1992b). Indeed, goldsinny were found to be the fastest recruiting species after a toxic algal bloom in Norway (Johannessen & Gjosaeter, 1990). However, it is clear that the future growth of the use of wrasse as cleaner-fish will greatly depend on the availability of wrasse greater than 10 cm length. A wrasse rearing facility that could produce certified disease-free wrasse would prove to be beneficial and would alleviate the pressure currently being imposed on wild stocks (Deady *et al.*, 1995). If exploitation of wild stocks continues, the development of a fishery management strategy is required in order to prevent the over-exploitation of wrasse stocks in areas close to salmon farms, and to ensure that sufficient standing stocks are maintained. Such a strategy should consider geographical variation in the biology of wrasse, which influences response of a population to exploitation. For quantitative fishery models to be developed, a greater knowledge of natural recruitment and mortality rates is needed. A long-term assessment of the population dynamics of a wild wrasse population, before and after exploitation by a wrasse fishery, might provide the information required for the development of an effective model.

Acknowledgements

This work was funded by the Commission of the European Communities Fisheries and Aquaculture Research Programme, contract number AQ.2.502. We wish to thank Siobhan Lysaght, Will Darwall and Mark Costello for helpful suggestions and we are sincerely grateful to Dr Paul Connolly, Dr Dan Minchin and Dr Michael Keating for advice and helpful discussion.

References

Bjordal, Å. (1988) Cleaning symbiosis between wrasses (Labridae) and lice infested salmon (*Salmo salar*) in mariculture. *International Council for the Exploration of the Sea, Mariculture Committee* 1988/F: **17** 8 pp.

Bjordal, Å. (1990) Sea lice infestation of farmed salmon: possible use of cleaner fish as an alternative method for delousing. *Canadian Technological Report of Fisheries and Aquatic Sciences* **1761**: 85–9.

Bjordal, Å. (1992) Cleaning symbiosis as an alternative to chemical control of sea lice infestation of Atlantic salmon. In *The Importance of Feeding Behaviour for the Efficient Culture of Salmonid Fishes* (Ed. by Thorpe, J.E. and Huntingford, F.A.). World Aquaculture Workshops, No. 4, World Aquaculture Society, pp. 53–60.

Bjordal, Å. (1993) A new pot design for capture of wrasse. *International Council for the Exploration of the Sea, Fishery and Fish Behaviour Working Group Meeting*, Gothenburg, Sweden, 19–20 April 1993, 2 pp.

Chan, T.-Y. and Ribbink, A.J. (1990) Alternative reproductive behaviour in fishes, with particular reference to *Lepomis macrochira* and *Pseudocrenilabrus philander*. *Environmental Biology of Fishes* **28**: 249–56.

Cheetham, C. (1981) *A survey of the larval teleosts in the inner Galway Bay area.* Unpublished B.Sc. thesis. University College, Galway.

Costello, M.J. and Bjordal, Å. (1990) How good is this natural control of sea lice? *Fish Farmer* **13**(3): 44–6.

Darwall, W.R.T., Costello, M.J. and Lysaght, S. (1992a) Wrasse–how well do they work? *Aquaculture Ireland* **50**: 26–9.

Darwall, W.R.T., Costello, M.J., Donnelly, R. and Lysaght, S. (1992b) Implications of life-history strategies for a new wrasse fishery. *Journal of Fish Biology* **41** (Supplement B): 111–23.

Deady, S. and Fives, J.M. (1995) The diet of corkwing wrasse *Crenilabrus melops* in Galway Bay, Ireland and in Dinard, Brittany, France. *Journal of the Marine Biological Association of the UK* **75**: 635–49.

Deady, S., Varian, S.J.A. and Fives, J.M. (1995) The use of cleaner fish to control sea lice on two Irish salmon (*Salmo salar*) farms with particular reference to wrasse behaviour in salmon cages. *Aquaculture* **131**: 73–90.

Dipper, F.A. (1976) *Reproductive biology of Manx Labridae.* Unpublished Ph.D. thesis, University of Liverpool.

Dipper, F.A. (1981) The strange sex lives of British wrasse. *New Scientist* **90**: 444–5.

Dipper, F.A., Bridges, C.R. and Menz, A. (1977) Age, growth and feeding in the ballan wrasse *Labrus bergylta* Ascanius, 1767. *Journal of Fish Biology* **11**: 105–20.

Dipper, F.A. and Pullin, R.S.V. (1979) Gonochorism and sex inversion in British Labridae. *Journal of Zoology (London)* **187**: 97–112.

Doyle, M.J. (1988) *An ecological study of fish larvae in the plankton of coastal waters, west of Ireland.* Unpublished Ph.D. thesis. National University of Ireland

Dunne, J.J. (1970) *A survey of the larval stages of the teleosts of Galway Bay and adjacent areas with notes on the biology of some of the adults.* Unpublished B.Sc. Thesis. University College, Galway.

Dunne, J.J. (1972) *A survey of the teleost larval stages occurring in the plankton of outer Galway Bay.* Unpublished M.Sc. thesis. National University of Ireland.

Fiedler, K. (1964) Verhaltensstudien an Lippfischen der Gattung. *Crenilabrus* (Labridae, Perciformes). *Zeitschrift fur Tierpsychologie* **21**: 6–68.

Fives, J.M. (1967) *A survey of the fish larval population in the plankton of the west of Ireland.* Unpublished Ph.D. thesis. National University of Ireland.

Fives, J.M. (1970) Investigations of the plankton of the west coast of Ireland–IV. Larval and postlarval stages of fishes taken from the plankton of the west coast in surveys during the years 1958–1966. *Proceedings of the Royal Irish Academy* **70B**: 15–93.

Fives, J.M. and O'Brien, F.I. (1976) Larval and post-larval stages of fishes recorded from the plankton of Galway Bay, 1972–3. *Journal of the Marine Biological Association of the UK* **56**: 197–11.

Hilldén, N.-O. (1978) An age-length key for *Ctenolabrus rupestris* (L.) in Swedish waters. *Journal du Conseil international pour l'Exploration de la Mer* **38**: 271–2.

Hilldén, N.-O. (1981) Territoriality and reproductive behaviour in the goldsinny, *Ctenolabrus rupestris* (L.) (Pisces, Labridae). *Behavioural Processes* **6**: 207–21.

Johannessen, T. and Gjøsaeter, J. (1990) Algeoppblomstringen i Skagerrak i Mai 1988–ettervirkninger på fisk og bunnfauna langs Sörlandskysten. *Flödevigen Meldinger* **6**: 87 pp.

Kennedy, M. and Fitzmaurice, P. (1969) Pelagic eggs and young stages of fishes taken on the south coast of Ireland in 1967. *Irish Fisheries Investigations Series B* **5**: 5–36.

Kennedy, M., Fitzmaurice, P. and Champ, T. (1973) Pelagic eggs of fishes taken on the Irish coast. *Irish Fisheries Investigations Series B* **8**: 1–23.

Kodricbrown, A. (1988) Effects of sex-ratio manipulation on territoriality and spawning success of male pupfish, *Cyprinodon pecosensis*. *Animal Behaviour* **36**: 1136–44.

Munoz, J.C. (1991) Manila Bay: Status of its fisheries and management. *Marine Pollution Bulletin* **23**: 311–14.

Muus, B.J. and Dahlstrom, P. (1974) *Guide to the Sea Fishes of Britain and North-Western Europe*. Collins, London.

Potts, G.W. (1985) The nest structure of the corkwing wrasse, *Crenilabrus melops* (Labridae: Teleostei). *Journal of the Marine Biological Association UK* **65**: 531–546

Quignard, J.-P. (1966) Recherches sue les Labridae (Poissons, Teleostéens, Perciformes) des côtes européennes: systématique et biologie. *Naturalia Monspeliensia (Zoologie)* **5**: 7–248.

Richards, L.J. and Schnute, J.T. (1992) Statistical models for estimating CPUE from catch and effort data. *Canadian Journal of Fisheries and Aquatic Sciences* **49**: 1315–27.

Ricker, W.E. (1958) Handbook of computations for biological statistics of fish populations. *Bulletin of the Fisheries Research Board of Canada* **119**: 300 pp.

Sayer, M.D.J., Cameron, K.S. and Wilkinson, G. (1994) Fish species found in the rocky sublittoral during winter months as revealed by the underwater application of the anaesthetic quinaldine. *Journal of Fish Biology* **44**: 351–3.

Sayer, M.D.J., Gibson, R.N. and Atkinson, R.J.A. (1993) Distribution and density of populations of goldsinny wrasse (*Ctenolabrus rupestris*) on the west coast of Scotland. *Journal of Fish Biology* **43** (Supplement A): 157–67.

Sayer, M.D.J., Gibson, R.N. and Atkinson, R.J.A. (1995) Growth, diet and condition of goldsinny on the west coast of Scotland. *Journal of Fish Biology* **46**: 317–40.

Sjölander, S., Larson, H.O. and Engström, J. (1972) On the reproductive behaviour of two labrid fishes, the ballan wrasse (*Labrus bergylta*) and Jago's goldsinny (*Ctenolabrus rupestris*). *Revue Comportement Animale* **6**: 43–51.

Treasurer, J.W. and Henderson, G. (1992) Wrasse: a new fishery. *World Fishing* **41**: 2.

Treasurer, J. (1994a) Prey selection and daily food consumption by a cleaner fish, *Ctenolabrus rupestris* (L.) on farmed Atlantic salmon, *Salmo salar* L. *Aquaculture* **122**: 269–77.

Treasurer, J.W. (1994b) Distribution and species length composition of wrasse (Labridae) in inshore waters of West Scotland. *Glasgow Naturalist* **22**: 409–17.

Treasurer, J.W. (1994c) The distribution, age and growth of wrasse (Labridae) in inshore waters of west Scotland. *Journal of Fish Biology* **44**: 905–18.

Tully, O. (1986) *The Bathyhyponeuston and Ichthyoneuston of Galway Bay*. Unpublished Ph.D. thesis. National University of Ireland.

Turner, G.F. (1993) Teleost mating behaviour. In *Behaviour of Teleost Fishes* 2nd edn. (Ed. by Pitcher, T.J.). Chapman and Hall, London, pp. 307–26.

Turner, J.R. and Warman, C.G. (1991) The mobile fauna of sublittoral cliffs. In *The Ecology of Lough Hyne* (Ed. by Myers, A.A., Little, C., Costello, M.J. and Partridge, J.C.). Royal Irish Academy, Dublin, pp. 127–38.

Van der Veer, H.W., Witte, J.I.J., Beumkes, H.A., Dapper, R., Jongejan, W.P. and Van der Meer, J. (1992) Intertidal fish traps as a tool to study long-term trends in juvenile flatfish populations. *Netherlands Jounal of Sea Research* **29**: 119–26.

Chapter 10
Survival, osmoregulation and oxygen consumption of wrasse at low salinity and/or low temperature

M.D.J. SAYER[1], J.P. READER[2] and J. DAVENPORT[3] [1] *Centre for Coastal and Marine Sciences, Dunstaffnage Marine Laboratory, P.O. Box 3, Oban, Argyll PA34 4AD, UK;* [2] *Department of Life Science, University of Nottingham, University Park, Nottingham NG7 2RD, UK; and* [3] *University Marine Biological Station Millport, Isle of Cumbrae KA28 0EG, UK*

The successful use of wrasse as cleaner-fish on salmon farms is compromised in part by large winter mortalities in cages. A possible method of enhancing survival by maintaining wrasse at high densities in land-based tanks with no refuges was investigated for goldsinny and rock cook. The tanks were subject to ambient water conditions (minimum temperature and salinity 5.1°C and 21.0‰ respectively) and, after 152 days, mean goldsinny survival was 65.2%, whereas only 8.0% of rock cook survived. The osmoregulation of rock cook, corkwing and goldsinny, and oxygen consumption of goldsinny, were examined during and after rapid reductions in temperature and/or salinity. All three species survived exposure to rapid declines in salinity down to 8‰ at temperatures of 6 and 8°C. Rock cook and corkwing had elevated plasma osmolality after 3 days' exposure to full-strength sea water at 4°C, and their survival was low (53.3 and 46.2% respectively) after salinity reduction (8, 16, 24‰) or maintenance at 32‰. Goldsinny survival was higher at 4°C irrespective of salinity (95.5%) and showed no evidence of impaired osmoregulation at low temperature. Below 6°C goldsinny entered a hypometabolic state, possibly an aid to survival over winter.

Goldsinny appear to be better adapted to winter conditions than rock cook or corkwing. Methods for improving wrasse survival in salmon cages over winter are discussed.

10.1   Introduction

Three species of wrasse in north Europe are commonly used to control infestations of sea lice in farmed Atlantic salmon (*Salmo salar* L.); goldsinny [*Ctenolabrus rupestris* (L.)], rock cook [*Centrolabrus exoletus* (L.)] and corkwing [*Crenilabrus melops* (L.)] (reviews: Bjordal, 1991; Costello, 1993; Treasurer, 1993). The environmental and financial benefits of using cleaner-fish are compromised, in part, by large losses of wrasse on some salmon farms over winter, necessitating restocking in spring (Kvenseth, 1993). The reasons for this winter loss are unclear, but are thought to result primarily from an inability to cope with low temperatures and low salinity (Treasurer, 1993). In areas populated by wrasse, salinities and temperatures as low as 12‰ and 5°C have been recorded at depths of 4–6 m, where episodic discharge of fresh water into sea lochs is having an influence (Sayer *et al.*, 1993; Treasurer, 1994). In the wild, the behaviour of the wrasse may allow them to survive these conditions. During winter, rock cook withdraw into deep crevices and become inactive (Sayer *et al.*, 1994). The behaviour of goldsinny is similar, although a certain amount of activity is always observed (Sayer *et al.*, 1993), and they continue to eat, albeit infrequently (Sayer *et al.*, 1995). However, corkwing seem to be more active, as they can be captured in creels or trawls throughout the winter, often in shallow areas (<10 m) strongly affected by fresh-water runoff (Sayer *et al.*, 1993; this volume, Chapter 2). There is no evidence for an offshore migration by any of the three species.

In salmon cages, there can be large fluctuations in temperature and salinity caused by freshwater run-off. The relatively shallow deployment of the cages, and the mixing of the water by the swimming action of the salmon, ensure that these changes affect the whole of the water enclosed by the cages. Decreases in temperature and salinity from 10 to 5°C and 27 to 8‰ respectively have been recorded over 48-hour periods at depths greater than 4 m in salmon cages (Treasurer, *unpubl. data*). Wrasse in salmon cages may be unable to cope with such conditions because they cannot take the avoiding action seen in the wild.

This account summarizes a series of experiments which was designed to investigate the tolerance of wrasse at low salinity and/or low temperature, and the effects on their physiology. Tolerance of rapid (5–6 days; goldsinny, rock cook, corkwing) and chronic (146 days; goldsinny, rock cook) temperature and salinity reductions was investigated. As well as testing long-term survival, the chronic tolerance test was designed to examine the possibility that winter survival could be enhanced in salmon farms by maintaining wrasse in an environment less challenging than that found in salmon cages. It was assumed that many farms could have shore-based tanks available for winter wrasse maintenance, but that temperature and salinity fluctuations would be similar to natural conditions (i.e., there would be no access to temperature or salinity control). A second assumption was that it would be financially beneficial to reduce husbandry requirements to a minimum, and so no refuges were provided, and high densities stocked. Finally, it may be necessary to maintain two or more wrasse species in the same tank. The object of the physiological studies was to determine the ability of goldsinny, rock cook and corkwing to osmoregulate under conditions of rapid reductions in temperature and salinity which they may encounter in salmon cages, and to investigate the metabolism of goldsinny at low temperature. Goldsinny have been observed in a torpid state at low temperature in the field (Costello, 1991; Sayer *et al.*, 1994).

If the survival of wrasse in salmon farms could be improved, exploitation of wild populations could be reduced, the use of hatchery-reared (and therefore disease-free) wrasse could become a commercial possibility, and the use of these cleaner-fish in place of pesticides could become a more widespread practice. The pesticides at present in use in the salmon industry are expensive both financially and in terms of their environmental impact.

10.2   Materials and methods

For the long-term salinity/temperature reduction test goldsinny ($n = 250$) and rock cook ($n = 350$) were donated by a commercial supplier (Lamont Wrasse) and immediately introduced into the experimental apparatus. Goldsinny, rock cook and corkwing wrasse were obtained 1–3 weeks before experiments testing survival and osmoregulation during rapid reductions in temperature and salinity. Goldsinny were captured using baited creel (Treasurer, 1991, for methodology); some corkwing were caught by beam trawling. Additional corkwing, and all rock cook were purchased from a commercial supplier. Prior to their use the wrasse were maintained in through-flow holding aquaria and fed daily on frozen mysid shrimp (Mysidacae). Wrasse were starved for 24 hours before introduction into the apparatus. The goldsinny required for measuring metabolic activity at low temperatures were captured using baited traps four days before initiation of each experiment. They were maintained in a clean, aerated, through-flow aquarium with adequate refuge provision. They were not fed between capture and use in order to minimize variation in metabolic rate associated with specific dynamic action (Jobling & Davies, 1980; Duthie & Houlihan, 1982; Jobling, 1993).

Long-term survival was tested in two 1 m dia. through-flow tanks each supplied with seawater at 2.4–3.0 l min^{-1}. Goldsinny and rock cook were stocked in each tank at densities of 159 and 223 m^{-3} respectively (total wrasse density per tank was 382 m^{-3}), with no refuge provision. On most days they were fed on frozen mysid, and the temperature and salinity of each tank was measured. At the same time, mortalities were removed and recorded. The experimental duration was 152 days (18 November 1993 to 19 April 1994).

The experimental apparatus for studying survival and osmoregulation of wrasse exposed to rapid reductions in temperature and salinity were located in three temperature-controlled rooms maintained at nominal temperatures of 8, 6 or 4°C. Each room contained twelve shaded Perspex tanks (capacity 12 l), to allow the exposure of each of the three species to one of four nominal salinities (32, 24, 16 or 8‰). Each tank was aerated. All rooms were equipped with continuous (24 hr) diffused, subdued artificial illumination. Wrasse were introduced into the tanks, which were filled initially with sea water of the same composition as in the holding aquaria (temperature, 14.3 ± 0.1°C; salinity, 31.3 ± 0.1‰; mean ± SE, $n = 15$), at densities of 6–7 per tank. The temperature was then reduced to target values over 5–10 hr; salinity was unchanged. All tanks were then maintained at 8, 6 or

4°C and salinity *c*. 32‰ for 69.5–78 hr, after which five to six fish of each species at each temperature (i.e., one to two fish from each experimental tank) were removed at random for blood-sampling. At each of the three temperatures, one tank of each species then continued to be maintained at nominal salinity 32‰. The salinity of the remaining three tanks of each species at each temperature was reduced in steps (*c*. 12 hr), over a period of some 36 hr, to a nominal salinity of either 24, 16 or 8‰. These salinities were then maintained approximately constant for a further 24 hr. At the end of this period, all surviving animals were removed for blood-sampling. Throughout the period of the experiment, most of the water in each tank was renewed with water of the appropriate salinity and temperature at intervals of *c*. 12 hr. Temperature and temperature-corrected salinity measurements were made after each water change. Mortalities were recorded and removed at regular intervals. A second experiment was carried out immediately after the first, using an identical protocol except that only corkwing and rock cook were used.

Animals for blood-sampling were killed by prolonged anaesthesia (ethyl p-amino benzoate; Sigma), blotted dry and weighed. Blood samples were removed by caudal amputation, treated with heparin and centrifuged for 10 min at 11 000 g. Blood plasma osmolality was determined by vapour pressure osmometry (Wescor: model 5500). For comparison purposes, 6–7 specimens of each species from the holding aquaria were sampled at the beginning of the experiments. The short-term survival and osmoregulation experiments were undertaken during August 1994.

Oxygen uptake was chosen as an indirect method of assessing goldsinny metabolic physiology at low temperatures. Goldsinny were held individually in opaque respirometers fitted with internal mixing devices. Grids fitted at the top of each respirometer allowed free movement of aerated water in and out of the chamber, but minimized disturbance to the fish from external movement. Three experimental chambers were held in a 12-l, seawater-filled, plastic aquarium, which was positioned in a temperature-controlled bath. Fish were introduced to the chambers at the water temperature measured in the pre-experimental through-flow maintenance aquarium (temperature range 11.7–13.4°C, *n* = 32). The water temperature was subsequently reduced to a nominal value of 10°C over four to five hours and maintained at 10°C for a further eight to ten hours prior to any oxygen-uptake recordings being made. Measurements were taken by securing a lid with integral O-ring into which was mounted an oxygen electrode (Radiometer E5046, connected to a Radiometer PHM71 Mk2 acid-base analyser and Rikadenki R-22 chart recorder) on each chamber in turn, ensuring that no air bubbles were trapped (aquarium aeration was temporarily suspended). The secured, whole chamber was moved from the seawater aquarium to the temperature-controlled bath and located on a submersible magnetic stirrer. A narrow ring attached to the base of each experimental chamber enabled correct location on the stirrer to ensure total mixing for the duration of measurement. The enclosed chamber was analysed for oxygen content until *c.* 15–20% of the available oxygen had been utilized (duration 21.0–43.2 minutes, *n* = 191). At 10°C, measurements were made until constant uptake slopes were obtained for each chamber; an average of 3.9 measurements were made on each fish at 10°C. Further measurements were then made at the other nominal temperatures of 8, 6 and 4°C; an average of 2.9 measurements were taken for each fish at each of these three temperatures. Each temperature was maintained for 14 hr; each 2°C temperature drop was achieved over 4 hr. Measurements of the temperature and salinity of the seawater within the experimental aquarium were made before each oxygen uptake recording. The water within the experimental aquarium was replaced at 12-hour intervals with fully aerated, temperature-equilibrated seawater. On termination of each experiment, fish were blotted dry and weighed. A fish density of $1 \, g \, ml^{-1}$ was assumed in calculating water volume. Oxygen solubility in sea water was calculated for each temperature/salinity combination using the nomogram of Green and Carritt (1967).

Prior to each experiment, oxygen uptake measurements were made on empty chambers which had previously been soaked in sodium hypochlorite solution. On termination of each experiment, oxygen uptake measurements were made on each chamber immediately following fish removal. A linear relationship for oxygen uptake between clean and dirty chambers was assumed, and temporally-corrected background values were deleted from total measurements. The oxygen electrode was calibrated at each temperature using nitrogen and air-saturated seawater. Four experimental replicates, using 12 goldsinny, were carried out with a decreasing temperature profile as outlined above; one run using three goldsinny was made over a similar duration, but at a constant nominal water temperature of 10°C. The oxygen consumption experiments were undertaken in July 1994.

10.3   Results

10.3.1   *Long-term survival*

In the long-term survival experiment, the temperature and salinity maxima and minima were 11.3 and 5.1°C, 32.8 and 21.0‰ respectively (Fig. 10.1a and b). The temperature averaged (±SE) 8.7 ± 0.2°C for the first 50 days (*n* = 50), 6.6 ± 0.1°C for the second 50 days (*n* = 40), and 6.5 ± 0.1 for the last 52 days (*n* = 45) (Fig. 10.1a). The average (±SE) salinities for the same periods were 31.1 ± 0.2‰ (*n* = 50), 29.9 ± 0.4‰ (*n* = 41) and 28.1 ± 0.4‰ (*n* = 45) (Fig. 10.1b). Most temperature fluctuations were ±0.6°C in 24 hr, although a temperature reduction of 2.9°C (9.3–6.4°C) was recorded over 48 hr (days 32–34; Fig. 10.1a). There were three main periods of low salinity, between days 31 and 38 (minimum 27.4‰), days 59–78 (minimum 22.8‰), and days 110–133 (minimum 21.2‰) (Fig. 10.1b).

Mean rock cook survival in the two tanks over the 152 days was 8.0% (11.4 and

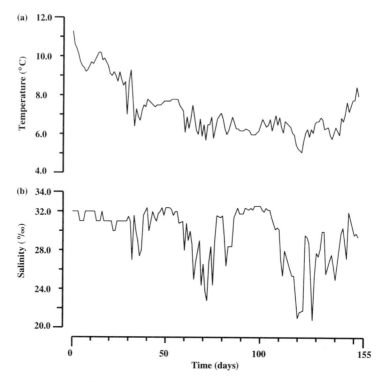

Fig. 10.1 Temperature and salinity profiles recorded over 152 days. Data pooled from the two experimental tanks.

4.6%); mean goldsinny survival was 65.2% (68.0 and 62.4%). The main mortality periods differed between the replicates, for both species (Fig. 10.2). The cumulative mortality of rock cook was only 6.3% in tank R2 on day 77 but then increased linearly after that to 90.9% on day 146 (Fig. 10.2a). Large numbers of rock cook mortalities were recorded from day 38 onwards in tank R1 and had reached 76.0% by day 77, and 88.6% on day 146 (Fig. 10.2a). In tank G2 mortality was low up to day 131; 50% of all mortalities in that tank were recorded in the last 21 days of the trial (Fig. 10.2b). Nearly all goldsinny mortalities in G1 were recorded between days 48 and 70 (Fig. 10.2b).

10.3.2   *Short-term survival*

Water quality parameters for the short-term survival experiments are given in Table 10.1. During the three-day period of temperature acclimation in full-strength (*c.* 32‰) seawater, rock cook and corkwing survival was 100% at 8 and 6°C; survival was 75.9 and 71.7% respectively at 4°C (Fig. 10.3). One goldsinny death occurred at 6°C, and none at 4 and 8°C (Fig. 10.3). Survival of all three wrasse species was high at 8 and 6°C, during and after salinity reduction (Fig. 10.4). One

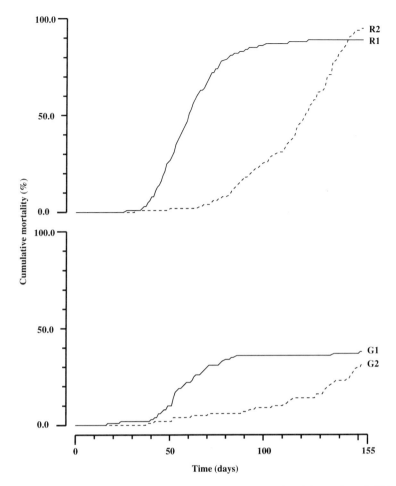

Fig. 10.2 Cumulative percentage mortality of replicate experiments involving rock cook (R1 and R2) and goldsinny (G1 and G2) exposed to winter salinity and temperature conditions for 152 days.

goldsinny died at 4°C during salinity reduction. Mortalities of corkwing and rock cook continued to occur during and after salinity reduction at 4°C, but there is little evidence to suggest that low salinity *per se* had any effect on survival. Survival data for the total experimental duration are given in Table 10.2.

10.3.3 *Plasma osmolality*

Blood plasma osmolalities (mean ± SE) of fish sampled from the holding aquaria before temperature and salinity reductions were $353.0 \pm 4.1 \, \text{mOsmol kg}^{-1}$ ($n = 6$) in rock cook, $342.3 \pm 3.8 \, \text{mOsmol kg}^{-1}$ ($n = 6$) in corkwing and $349.6 \pm 1.7 \, \text{mOsmol kg}^{-1}$ ($n = 7$) in goldsinny.

At 4°C, there was a pronounced rise in plasma osmolality of corkwing and rock

Table 10.1 Water quality for short-term survival and osmo-regulation experiments; mean ± SE.

Species	Nominal temperature (°C)	Temperature acclimation (69.5–78 hr)		Actual temperature (°C) (n = 24–53)	Actual salinity (‰) (nominal 32‰) (n = 6–10)	Salinity reduction (Salinities measured after reduction)		
		Actual temperature (°C) (n = 20–48)	Actual salinity (‰) (n = 15–34)			Actual salinity (‰) (nominal 24‰) (n = 3–8)	Actual salinity (‰) (nominal 16‰) (n = 3–7)	Actual salinity (‰) (nominal 8‰) (n = 3–8)
Goldsinny	8	8.0 ± 0.0	31.4 ± 0.1	8.1 ± 0.0	31.5 ± 0.1	24.2 ± 0.3	16.5 ± 0.4	9.1 ± 0.6
	6	6.4 ± 0.1	31.6 ± 0.0	6.5 ± 0.1	31.4 ± 0.0	23.8 ± 0.2	17.4 ± 0.4	8.3 ± 0.7
	4	4.6 ± 0.1	31.5 ± 0.0	4.6 ± 0.1	31.5 ± 0.1	23.6 ± 0.5	15.7 ± 0.4	7.7 ± 0.1
Rock cook	8	8.1 ± 0.0	31.4 ± 0.0	8.1 ± 0.0	31.3 ± 0.1	24.6 ± 0.2	17.0 ± 0.3	8.8 ± 0.2
	6	6.4 ± 0.0	31.5 ± 0.0	6.4 ± 0.0	31.2 ± 0.1	24.1 ± 0.2	16.8 ± 0.4	8.5 ± 0.4
	4	4.5 ± 0.0	31.5 ± 0.0	4.5 ± 0.0	31.4 ± 0.2	24.9 ± 0.6	16.2 ± 0.2	8.2 ± 0.1
Corkwing	8	8.0 ± 0.0	31.5 ± 0.0	8.0 ± 0.0	31.3 ± 0.1	24.7 ± 0.3	16.9 ± 0.3	8.4 ± 0.2
	6	6.4 ± 0.0	31.5 ± 0.0	6.4 ± 0.0	31.3 ± 0.1	24.3 ± 0.3	16.5 ± 0.2	8.2 ± 0.5
	4	4.6 ± 0.0	31.4 ± 0.0	4.5 ± 0.0	31.3 ± 0.1	24.8 ± 0.3	16.6 ± 0.3	8.0 ± 0.0

Table 10.2   Total percentage mortality of rock cook, corkwing and goldsinny before, during and after exposure to nominal salinities at temperatures of 8, 6 or 4°C.

Species	Temperature (°C)	Nominal salinity			
		32‰	24‰	16‰	8‰
Rock cook	8	0.0	0.0	0.0	9.1
	6	0.0	0.0	0.0	10.0
	4	63.6	50.0	50.0	83.3
Corkwing	8	0.0	0.0	0.0	16.7
	6	0.0	0.0	0.0	0.0
	4	45.5	66.7	88.9	88.9
Goldsinny	8	0.0	0.0	0.0	9.1
	6	0.0	0.0	0.0	0.0
	4	0.0	0.0	16.7	0.0

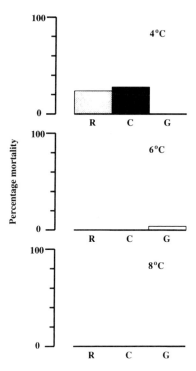

Fig. 10.3   Percentage mortality of rock cook (R), corkwing (C) and goldsinny (G) acclimated to 4, 6 or 8°C for three days.

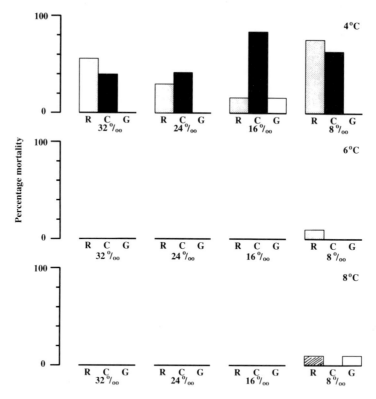

Fig. 10.4 Total percentage mortality of rock cook (R), corkwing (C) and goldsinny (G) exposed to nominal salinities of 32, 24, 16 and 8‰ at nominal temperatures of 4, 6 and 8°C.

cook after 72 hr at 32‰ salinity (significant difference from t = 0 values: $P <$ 0.001 in each case; Student's *t*-test). This rise was not observed in these species at 6 or 8°C, nor in goldsinny at any temperature. Indeed, plasma osmolality of goldsinny tended to fall slightly after 72 hr, significantly so at 6°C ($P < 0.05$) (Fig. 10.5). At the end of the experiment, plasma osmolalities of all three species at 32‰ salinity were similar to those at t = 0 ($P > 0.05$ where sample numbers were large enough to allow significance testing) (Figs 10.5 and 10.6).

At the end of the experiment, plasma osmolalities of all three species showed evidence of a slight decline with decline in salinity (Fig. 10.6). At 24‰ salinity there were no differences from t = 0 values ($P > 0.05$ where tests were possible). Osmolality of corkwing and rock cook at 16 and 8‰ tended to be lower than at t = 0 (of the results from 8 and 6°C, only corkwing at 16‰ and 6°C did not show a significant difference at the $P < 0.05$ level). Osmolality of goldsinny at 8‰ was significantly lower than at t = 0, at 8 and 6°C ($P < 0.01$) but not at 4°C ($P > 0.05$).

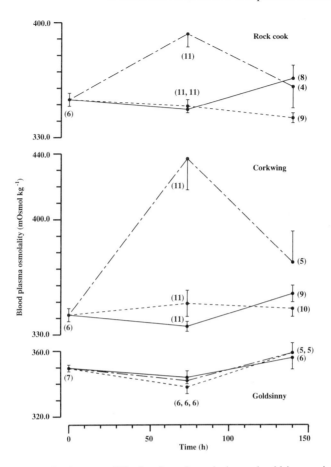

Fig. 10.5 Plasma osmolality (mean ± SE) of rock cook, corkwing and goldsinny maintained in 32‰ seawater at 8 (●——●), 6 (●-----●) or 4°C (●—·—·—●) for *c*. 140 hr. Sample sizes given in parentheses.

10.3.4 *Oxygen consumption*

Water quality measurements for the oxygen consumption experiments are given in Table 10.3. Mean oxygen uptake rates (±SE) fell from 0.042 (±0.003) to 0.034 (±0.005) ml $O_2 g^{-1} h^{-1}$ at 10 and 8°C respectively (Q_{10} = 2.86) and to 0.028 (±0.006) ml $O_2 g^{-1} h^{-1}$ at 6°C (Q_{10} = 2.58) (Fig. 10.7). The Q_{10} value between 10 and 6°C was 2.71. Between 6 and 4°C mean rates decreased to 0.008 (±0.002) ml $O_2 g^{-1} h^{-1}$ (Q_{10} = 542.01) (Fig. 10.7). The rate of oxygen uptake at 4°C was significantly lower than rates measured at 10, 8 and 6°C (Student's *t*-test; $P < 0.001, 0.001, 0.01$ respectively). There were no obvious trends in uptake rates recorded for goldsinny maintained at a constant nominal 10°C for 3 days; mean uptake rate (±SE) for the three fish was $0.046 ± 0.003$ ml $O_2 g^{-1} h^{-1}$ (n = 28).

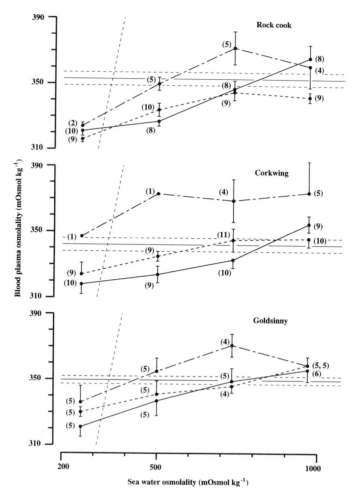

Fig. 10.6 Blood plasma osmolality (mean ± SE) of rock cook, corkwing and goldsinny maintained at 8 (●——●), 6 (●-----●) or 4°C (●—·—·—●) against measured seawater osmolality. Control plasma osmolalities denoted by horizontal solid (mean) and broken lines (±SE). Iso-osmotic relationship denoted by diagonal broken line. Sample sizes given in parentheses.

Table 10.3 Water quality in oxygen consumption experiments; mean ± SE.

Nominal temperature (°C)	Measured temperature (°C) (n = 44–53)	Measured salinity (‰) (n = 31–77)
10	9.90 ± 0.03	32.1 ± 0.1
8	7.96 ± 0.04	32.3 ± 0.1
6	5.83 ± 0.02	32.4 ± 0.2
4	3.90 ± 0.04	31.9 ± 0.1

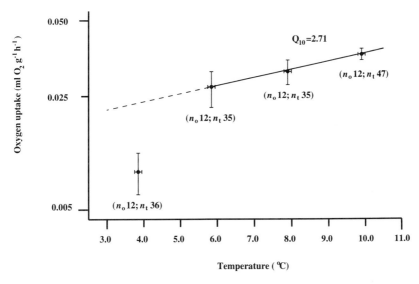

Fig. 10.7 Oxygen uptake (ml O_2 g^{-1} h^{-1}; mean ± SE, log$_{10}$-transformed) of goldsinny maintained at nominal temperatures 10, 8, 6 or 4°C (actual temperature: mean ± SE; sample numbers n_t). Oxygen uptake sample sizes given in parentheses (n_o), each sample being the mean of 2–4 replicate measurements per fish at each temperature. The solid and broken lines represent a linear dependence of oxygen uptake on temperature with a Q$_{10}$ of 2.71.

10.4   Discussion

There were distinct differences in the abilities of the three wrasse species to survive exposure to low temperatures. In the long-term survival study, about 65% goldsinny survived compared with only 8% rock cook. During rapid temperature reductions, goldsinny survival was 96% compared with 53 and 46% for rock cook and corkwing respectively. Low salinity appeared to have only a small effect on the survival of corkwing and rock cook, and none at all on the survival of goldsinny, in the short-term exposures. The water quality in both the long- and short-term survival experiments were relevant to shallow areas on the west coast of Scotland vulnerable to large and rapid discharges of freshwater from nearby large-catchment sea lochs (Sayer *et al.*, 1993). Goldsinny and rock cook are generally absent from areas or depths affected by low salinities (Sayer *et al.*, 1993; this volume, Chapter 2) but are found in the shallow rocky subtidal during the winter, and would be exposed to gradual declines in temperatures (Sayer *et al.*, 1994). Corkwing have been captured regularly throughout a winter in shallow water (2–5 m) in a bay directly influenced by periodic freshwater input (this volume, Chapter 2). Measurements in the water body made in a separate study indicate that corkwing inhabiting areas of this type may experience rapid temperature fluctuations (Sayer *et al.*, 1993). The results of the present study, therefore, do not readily accord with field observations. Large differences between goldsinny and corkwing survival (mean: 94 and 13% respectively) were also recorded in netcages submerged at depths of 17 and 25 m for six months over winter (this volume, Chapter 22), with a minimum temperature of 3.9°C. It is possible that increased depth (corkwing have been reported down to only 15 m, Quignard, 1966; Costello, 1991) may in part explain the high corkwing mortality, but the figures for goldsinny and corkwing in that study are comparable with those for the 4°C treatments in the short-term study.

From survival data, there is little evidence of vulnerability to low salinity in any of the three species: corkwing mortalities were slightly higher at low salinities and rock cook mortalities showed a tendency to be greatest at the extremes of salinity. Corkwing, and to a lesser extent rock cook, were showing evidence of failure to hypo-osmoregulate after three days at 4°C. It was shortly after the acclimation period, however, that the main mortalities began (most severe in corkwing), so the failure to regulate is possibly a reflection of the moribund state of many of the animals sampled at this time. Survivors at the end of the experiment, even at the lowest temperatures, showed evidence of quite effective hypo- and hyper-osmoregulation over the salinity range.

The greater ability of goldsinny to tolerate rapid salinity and temperature reductions compared with the other two wrasse species cannot be explained by a higher tolerance of low salinity. Quignard (1966) reported the resistance to low salinity (*c.* 7‰) of corkwing and goldsinny to be similar at 18–20°C. However, low temperature reduces the salinity tolerance of fish (Finstad *et al.*, 1988; Malloy & Targett, 1991). Consequently, a greater tolerance of low temperature may be of benefit in combating the effects of low salinity at low temperatures. Goldsinny have been observed in a torpid state in the field at low winter temperatures (Sayer *et al.*, 1994). Measurements of goldsinny oxygen consumption at low temperatures indicated that this species enters a hypo-metabolic state at temperatures below 6°C. In hyper-osmotic or hypo-osmotic media, the gill surface is a major site of passive movement of water and solutes down the concentration gradients (Evans, 1993). Passive ion fluxes are affected little by temperature ($Q_{10} = 1.2–1.4$) whereas ATP-dependent ion pumps are strongly affected ($Q_{10} = 2–4$) (Hochachka, 1988). If a fish is exposed to a temperature fall, there will tend to be a mismatch between passive and active ion fluxes (Davenport and Sayer, 1993). Failure to contend with this imbalance at low temperature in a hyperosmotic medium may explain the tendency for increased osmolality in the plasma of corkwing and rock cook after three days at 4°C.

Deep torpor in winter goldsinny has been observed in the field (Sayer *et al.*, 1994) but has not before been observed in the laboratory. The degree of torpor was so great in some animals at 4°C, and their oxygen uptake rates so low (below detection in some cases), that, on being removed from the respirometers, they appeared to be dead. Normal activity resumed in all cases, however, after a temperature increase of about 1°C, and no mortalities were recorded. Winter dormancy in fish has been reported previously (Walsh *et al.*, 1983; Crawshaw,

1984), but the scale of metabolic depression in those accounts (Q_{10} values 4.10 and *c*. 3.65 respectively) does not compare with the values recorded for goldsinny oxygen uptake rates ($Q_{10} = 542.01$). Hazel and Prosser (1974) postulate that a high Q_{10} at low temperatures may be highly adaptive for survival as it would promote metabolic depression and thus energy conservation during cold-induced dormancy, and facilitate a rapid increase in metabolism to a near-optimal rate in response to any temperature rise.

Goldsinny survival was relatively high when maintained in high-density conditions with no refuge provision, even though they were exposed to actual winter seawater conditions. The experimental design was purposely simple in order to mimic possible salmon farm requirements (e.g. no water treatment, minimal husbandry). It is possible that additional investment in refuge design may enhance survival considerably. In one tank (R2), rock cook survival was relatively high (>80%) for 93 days, during which temperature and salinity had fallen to 5.7°C and 24.8‰ respectively. The earlier mortalities (both rock cook and goldsinny) in the other tank (R1 and G1) suggest that death was caused by something other than low temperature or low salinity. It is possible that the later wrasse mortalities were caused by ambient water quality alone.

The behaviour of goldsinny and rock cook in winter in the wild, retreating deep into rocky scree (Sayer *et al.*, 1994), and the ability of goldsinny to enter a hypo-metabolic state at low temperatures, suggests that in these species sedentary behaviour is an adaptation for survival in winter. Corkwing, however, appear to remain active during the winter: perhaps they survive by responding to rapid temperature reductions by actively searching for areas of higher water temperature. Corkwing have been observed on rare occasions in deeper water during the winter (Sayer *et al.*, 1994). These interspecific differences in behaviour and temperature tolerances imply that the survival of goldsinny, and possibly rock cook, in winter in salmon farms could be enhanced if refuge design within the cages were changed in some way, to reduce disturbance when these fish are in an inactive state. Perhaps corkwing cannot ever be kept successfully in cages over winter, because they are being prevented from actively avoiding low temperatures.

Because of the difficulty of catching wrasse in winter, the short-term survival and osmoregulation experiments were undertaken during the summer months. It is possible that the high mortality of rock cook and corkwing at 4°C may have been the result of exposure of summer fish to winter conditions. Preliminary results from a small-scale experiment carried out on corkwing in March 1995 indicate that winter fish are better adapted to coping with winter conditions (Sayer & Reader, *unpubl. data*). Although not seasonally correct, the present study does indicate that corkwing and rock cook are more susceptible to stress than are goldsinny. This observation, in addition to the contrasting winter survival rates of corkwing and goldsinny reported by Bjelland *et al.* (Chapter 22, this volume), suggests that in terms of potential annual survival, goldsinny may be the most appropriate of the three wrasse species for use in salmon farms.

Acknowledgements

This study was funded by Argyll and the Islands Enterprise, the Scottish Salmon Growers Association and the Crown Estate.

References

Bjordal, Å. (1991) Wrasse as cleaner-fish for farmed salmon. *Progress in Underwater Science* **16**: 17–28.

Costello, M.J. (1991) Review of the biology of wrasse (Labridae: Pisces) in northern Europe. *Progress in Underwater Science* **16**: 29–51.

Costello, M.J. (1993) Review of methods to control sea lice (Caligidae: Crustacea) infestations on salmon (*Salmo salar*) farms. In *Pathogens of Wild and Farmed Fish: Sea Lice* (Ed. by Boxshall, G.A. and Defaye, D.) Ellis Horwood, Chichester, pp. 219–52.

Crawshaw, L.I. (1984) Low-temperature dormancy in fish. *American Journal of Physiology* **246**: R479–86.

Davenport, J. and Sayer, M.D.J. (1993) Physiological determinants of distribution in fish. *Journal of Fish Biology* **43** (Supplement A): 121–45.

Duthie, G. and Houlihan, D.F. (1982) The effect of single step and fluctuating temperature changes on the oxygen consumption of flounders, *Platichthys flesus* (L.): lack of temperature adaptation. *Journal of Fish Biology* **21**: 215–26.

Evans, D.H. (1993) Osmotic and ionic regulation. In *The Physiology of Fishes* (Ed. by Evans, D.H.) CRC Press, New York, USA, pp. 315–41.

Finstad, B., Staunes, M. and Reite, O.B. (1988) Effect of low temperature on sea-water tolerance in rainbow trout, *Salmo gairdneri. Aquaculture* **72**: 319–28.

Green, E.J. and Carritt, D.E. (1967) New tables for oxygen saturation of sea water. *Journal of Marine Research* **25**: 140–47.

Hazel, J.R. and Prosser, C.L. (1974) Molecular mechanisms of temperature compensation in poikilotherms. *Physiological Reviews* **54**: 622–70.

Hochachka, P.W. (1988) Channels and pumps–determinants of metabolic cold adaption strategies. *Comparative Biochemistry and Physiology* **90B**: 515–19.

Jobling, M. (1993) Bioenergetics: feed intake and energy partitioning. In *Fish Ecophysiology* (Ed. by Rankin, J.C. and Jensen, F.B.) Chapman and Hall, London, pp. 1–44.

Jobling, M. and Davies, P.S. (1980) Effects of feeding on the metabolic rate and the Specific Dynamic Rate in plaice, *Pleuronectes platessa* L. *Journal of Fish Biology* **16**: 629–38.

Kvenseth, P.G. (1993) Use of wrasse to control salmon lice. In *Fish Farming Technology* (Ed. by Reinertsen, H., Dahle, L.A., Jørgensen, L. and Tvinereim, K.) Balkema, Rotterdam, pp. 227–32.

Malloy, K.D. and Targett, T.E. (1991) Feeding, growth and survival of juvenile summer flounder *Paralichthys dentatus*: experimental analysis of the effects of temperature and salinity. *Marine Ecology Progress Series* **72**: 213–23.

Quignard, J.-P. (1966) Recherches sur les Labridae (Poissons, Teleostéens, Perciformes) des côtes européennes: systématique et biologie. *Naturalia Monspeliensia (Zoologie)* **5**: 7–248.

Sayer, M.D.J., Cameron, K.S. and Wilkinson, G. (1994) Fish species found in the rocky sublittoral during winter months as revealed by the underwater application of the anaesthetic quinaldine. *Journal of Fish Biology* **44**: 351–3.

Sayer, M.D.J., Gibson, R.N. and Atkinson, R.J.A. (1993) Distribution and density of populations of goldsinny wrasse (*Ctenolabrus rupestris*) on the west coast of Scotland. *Journal of Fish Biology* **43** (Supplement A): 157–67.

Sayer, M.D.J., Gibson, R.N. and Atkinson, R.J.A. (1995) Growth, diet and condition of goldsinny on the west coast of Scotland. *Journal of Fish Biology* **46**: 317–40.

Treasurer, J.W. (1991) Wrasse need due care and attention. *Fish Farmer* **14**(4): 24–6.

Treasurer, J.W. (1993) Management of sea lice (Caligidae) with wrasse (Labridae) on Atlantic salmon (*Salmo salar* L.) farms. In *Pathogens of Wild and Farmed Fish: Sea Lice* (Ed. by Boxshall, G.A. and Defaye, D.) Ellis Horwood, Chichester, pp. 335–45.

Treasurer, J.W. (1994) The distribution, age and growth of wrasse (Labridae) in inshore waters of west Scotland. *Journal of Fish Biology* **44**: 905–18.

Walsh, P.J., Foster, G.D. and Moon, T.W. (1983) The effects of temperature on metabolism of the American eel *Anguilla rostrata* (LeSueur): compensation in the summer and torpor in the winter. *Physiological Zoology* **56**: 532–40.

Chapter 11
Preliminary breeding trials of wrasse in an intensive system

A.B. SKIFTESVIK, K. BOXASPEN and A. PARSONS *Institute of Marine Research, Austevoll Aquaculture Research Station, N-5392 Storebø, Norway*

Broodstock of goldsinny, rock cook and corkwing wrasse were captured, and maintained in through-flow holding tanks at 12°C. Fish were stripped, and fertilized eggs incubated at 18°C in 10l containers. Larvae were held in similar containers and fed with algae and rotifers, and later *Artemia* sp. before weaning to dry feed. Best results were achieved for rock cook, where hatching occurred at approximately 90 degree-days, with considerable variation in survival beyond first-feeding. A high percentage of hatching was obtained with corkwing, although survival beyond day six was not as promising as that obtained for rock cook. High mortality was experienced at the egg stage of goldsinny with no survival beyond first-feeding. Based on experience from these trials, possible improvements for the holding of broodstock, and conditions for the different developmental stages are discussed. The experience gained during these trials indicates the requirement for further examination of holding conditions for broodstock and frequency of stripping.

11.1   Introduction

Infestation with sea lice, *Lepeophtheirus salmonis* Krøyer and *Caligus elongatus* Nordmann, is one of the major economic loss factors in commercial rearing of Atlantic salmon (*Salmo salar* L.) in the northern hemisphere. Lice eat mucus, skin and blood from the host, creating osmoregulatory problems and enhancing considerably the possibility of secondary infections (Brandal *et al.*, 1976; Pike, 1989; Johnson & Albright, 1992). Until recently, chemical treatment of whole farms was the only control for the lice problem (Brandal & Egidius, 1979; Boxaspen & Holm, 1991). The cleaning behaviour of wrasse has been used on a farming scale in Norway since 1991 following aquaria trials by Å. Bjordal (Bjordal, 1988; Costello & Bjordal, 1990; Skog, 1994). Farmers have based their practice of this method on wild wrasse caught in close proximity to farms. This has been recommended in order in minimize the risk of spreading diseases. In regions with few or no local populations of wrasse, stocks have been transported in from non-salmon farming areas. One problem with the use of wrasse is that they are difficult to catch before mid-June, towards the end of their spawning season. However, salmon farmers would ideally like to stock wrasse in the cages with the introduction of smolts in spring.

The problems of limited availability and health status could possibly be overcome by winter storage of fish in controlled systems (this volume, Chapter 22) or rearing of wrasse from broodstock. At Austevoll Aquaculture Research Station, intensive farming of halibut [*Hippoglossus hippoglossus* (L.)], turbot [*Scophthalmus maximus* (L.)] and (*Gadus morhua* L.) in semi-intensive and intensive systems has been carried out since 1985. These systems, and the knowledge required in their operation, have been used in a preliminary study into the rearing of wrasse.

11.2 Material and methods

Broodstock of rock cook [*Centrolabrus exoletus* (L.)], corkwing [*Crenilabrus melops* (L.)] and goldsinny [*Ctenolabrus rupestris* (L.)], were captured using fyke nets in June 1994 in the immediate area surrounding Austevoll Aquaculture Research Station, Norway. These were transferred to 75 l holding tanks supplied with sand-filtered sea water (6°C), from 50 m depth, at a rate of approximately 1.5 l per min. A constant room temperature of 23–25°C enabled the water temperature in the tanks to be maintained at 12°C. Numbers of females collected of each species were 35 corkwing, 40 rock cook and 35 goldsinny.

All fish were checked for ripe eggs upon first capture and subsequently at three to four-day intervals. Eggs were stripped manually, with rock cook and corkwing eggs being stripped onto small ($20 \times 10 \times 0.3$ cm) black polyethylene boards. Milt collected from males using a syringe, was added, and the eggs were left to stand with a covering of water for 5 min. to allow fertilization, and adhesion of the eggs to the boards. Opaque rock cook eggs were considered as overripe and were discarded. No such distinction was made with corkwing eggs. Goldsinny eggs were stripped into 75 ml containers and allowed to stand for 5 min. following the addition of milt and a small amount of water. Only floating eggs were incubated after this period as eggs that sank appeared opaque and were regarded as overripe.

Fertilized eggs were transferred to 10 l containers which were supplied with filtered seawater at 18°C at a rate of 75 ml per min. Each batch was placed in an individual container. Gentle aeration provided circulation in the containers. Eggs that died prior to hatching were removed from the containers by siphoning. Water temperature in each container was recorded daily.

After hatching, the polyethylene boards were removed from the containers and unhatched eggs removed. From day one, post-hatch, food was supplied to the tanks from a separate header tank system with a constant supply to each container of approximately 25 ml per min for up to 16 hours per day. Food consisted of a mixture of *Tetraselmis suecica* (Kylin) Butch and the portion of rotifers (*Brachionus plicatilis* Mueller) which passed through a 180 µm mesh sieve. Rotifers were gradually replaced by *Artemia* nauplii, enriched with Selco (Artemia Systems), from day 25 and *T. seucica* was discontinued.

Larvae of all species were filmed using a Wild M5 binocular microscope, without anaesthetic, and prints were made from a video sequence of the fish.

11.3   Results and discussion

11.3.1   *Goldsinny*

Goldsinny was the earliest of the three species to strip at the beginning of the season. The eggs were of a very variable quality both within and between batches. The percentage of overripe eggs at each stripping increased throughout the season to a situation where all eggs stripped were overripe. This problem could probably be solved by increasing the frequency of stripping.

Fertilized eggs developed satisfactorily through half the incubation time when large mortalities were observed and only a small fraction of the eggs reached hatching. The cause of the mortality at this stage in the incubation is not clear. Possible explanations could be the incubation system, the physical environment, bacterial infection, initial egg quality, or a combination of some or all of these. Of a total of nine groups, three gave hatching at about 67–73 day-degrees. Most larvae died after three days with only one group surviving to day six, post-hatch.

11.3.2   *Corkwing*

Corkwing produced the largest and best quality egg groups at the start of the experiment; quality was a subjective visual judgement of transparency and symmetry. The eggs were yellow and attached themselves within minutes to a surface. Overripe eggs were the possible cause of total mortality before hatching, as contact with dead eggs appeared to result in further mortality in the eggs surrounding them during the incubation period. Hatching in darkness gave better results than hatching in light suggesting that corkwing eggs hatch during darkness in the wild.

In the first part of the experiment, plates of white hard plastic were used as an attachment surface for the eggs. Even with good hatching, the larvae died within one day of hatching. They were observed to be strongly phototactic from the time of hatching and used a lot of energy constantly swimming against the white surface. Larvae hatched on grey-black surfaces distributed themselves evenly in the total water column of the incubators. Evenly distributed light was of vital importance as larvae were attracted to any source of light. Corkwing larvae were observed to graze on algal growth on the walls of the first-feeding tank. Of the 16 groups incubated, five groups hatched at 125 to 130 day-degrees. Most fish died five to seven days, post-hatch, with only one surviving past first-feeding.

11.3.3   *Rock cook*

Rock cook produced egg batches later in the year than corkwing and goldsinny, and also had a tendency to produce more overripe eggs. The fertilized eggs took longer to attach themselves than corkwing eggs. However, the time to attach seemed to be correlated to the quality of the eggs, with good quality eggs attaching faster, though this was not studied in detail. The newly hatched larvae were evenly distributed in the tank and did not seem to have the same strong positive phototactic response as corkwing. Out of 35 groups, eleven hatched and four gave larvae that survived past first-feeding. Hatching occurred at 85–95 day-degrees with most of the larvae dying five to seven days later. A total of 25–30 fish survived.

11.3.4   *Larvae appearance before first-feeding*

Larvae of corkwing show a distinct pattern of brown rings and blackish spots (Fig. 11.1) while rock cook only have the blackish spots in a very regular pattern on both sides of the body (Fig. 11.2a, dorsal view, 11.2b lateral view). The goldsinny however, has colouration only on the tail, gut and stomach areas (Fig. 11.3).

11.4   **Concluding remarks**

This preliminary experiment was conducted in order to obtain practical knowledge in stripping, egg incubation and handling of broodstock and larvae. The following conclusions can be made from our own experiment and which are supported by Stone (*pers. comm.*).

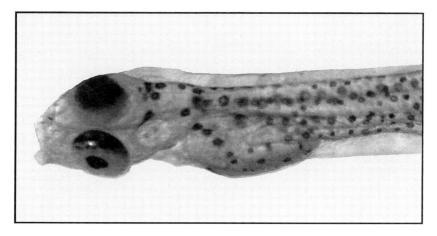

Fig. 11.1 Corkwing larvae before first-feeding. The pigmentation is a combination of brown-black rings and smaller spots.

(a)

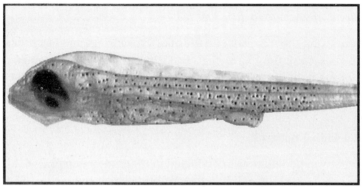

(b)

Fig. 11.2 (a) Dorsal view of rock cook before first-feeding. The pigmentation is a regular pattern of small black spots. (b) Lateral view of rock cook larvae where the regular distribution of spots can be seen along the body.

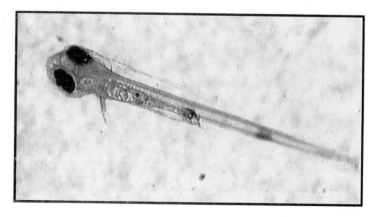

Fig. 11.3 Goldsinny larvae with distinct areas of pigmentation on the tail, gut and stomach.

The reduction of organic load in incubators is important in order to enhance survival. Thus, eggs should be washed after fertilization and before incubation. To reduce the bacterial load in the egg incubation period, the proportion of water volume to egg number should be higher for the goldsinny, which have pelagic eggs, and the flow increased for the corkwing and the rock cook which attach their eggs. This was not feasible in the present study because the supply of water of optimal temperature was limited. However stagnant water and, also darkness for corkwing, can be favourable just before and during the actual hatching process. Newly hatched larvae should be transferred to new containers in order to reduce bacterial load.

The broodstock of goldsinny should be stripped more frequently (probably every day) and the period of acclimation should be investigated. It is possible that egg quality was unstable for the year of capture. These conclusions are similar to those presented in Chapter 12 of this volume.

References

Bjordal, Å. (1988) Cleaning symbiosis between wrasses (Labridae) and lice infested salmon (*Salmo salar*) in mariculture. *International Council for the Exploration of the Sea, Mariculture Committee 1988/F* **17**: 8 pp.

Brandal, O.L. and Egidius, E. (1979) Treatment of salmon lice (*Lepeophtheirus salmonis*, Krøyer 1838) with Neguvon – description of method and equipment. *Aquaculture* **18**: 183–8.

Brandal, O.L., Egidius, E. and Romslo, I. (1976) Host blood: A major food component for the parasitic copepod *Lepeophtheirus salmonis* Krøyer, 1838 (Crustacea: Caligidae). *Norwegian Journal of Zoology* **24**: 341–3.

Boxaspen, K. and Holm, J.C. (1991) New biocides used against sea lice compared to organo-phosphorus compounds. In *Aquaculture and the Environment* (Ed. by de Pauw, N. and Joyce, J.). European Aquaculture Society Special Publication No. 16, pp. 393–402, Gent, Belgium.

Costello, M.J. and Bjordal, Å. (1990) How good is this natural control on sea lice? *Fish Farmer* **13**(3): 44–6.

Johnson, S.C. and Albright, L.J. (1992) Comparative susceptibility and histopathology of the response of naïve Atlantic, chinook and coho salmon to experimental infection with *Lepeophtheirus salmonis* (Copepoda: Caligidae). *Diseases of Aquatic Organisms* **14**: 179–93.

Pike, A.W. (1989) Sea lice – major pathogens of farmed atlantic salmon. *Parasitology Today* **5**: 291–7.

Skog, K. (1994) *Biological delousing of salmon with wrasse: optimizing the method with dietary and behavioural studies*. Unpublished thesis, Can scient., University of Bergen, Norway (in Norwegian).

Chapter 12
Preliminary trials on the culture of goldsinny and corkwing wrasse

J. STONE *Golden Sea Produce Ltd, Hunterston, West Kilbride, Ayrshire, UK (present address: Institute of Aquaculture, University of Stirling, Stirling, FK9 4LA, UK)*

From 1991 to 1992 approximately 5000 juvenile goldsinny [*Ctenolabrus rupestris* (L.)] and 1000 corkwing wrasse [*Crenilabrus melops* (L.)] were reared from eggs and milt collected by manual stripping of wild broodstock fish. Eggs were obtained both in the natural spawning season from fish held under ambient conditions and from out-of-season spawning by manipulation of temperature and photoperiod. Larvae were reared on cultures of a marine algae (*Isochrysis galbana* Parke) and, the rotifer (*Brachionus plicatilis* Mueller) or the copepod [*Eurytemora velox* (Lilljeborg)]. After 20–30 days post-hatch, nauplii and ongrown *Artemia salina* were introduced, following which the juveniles were gradually weaned onto non-live feeds. Survival from hatching to the juvenile stage in goldsinny wrasse was up to 7%, with individuals attaining a length of 9 cm in 12–15 months. Survival of corkwing wrasse was up to 32%, with individuals attaining 13 cm in 12–15 months. Initial results suggest wrasse may be suitable for rearing on a commercial scale with lengths of over 10 cm. They may also be suitable for stocking alongside Atlantic salmon, this being possible after approximately one year for corkwing and two years for goldsinny.

12.1   Introduction

Recent improvements in the control and treatment of furunculosis (*Aeromonas salmonicida*) have meant that sea lice (*Lepeophtheirus salmonis* Krøyer), an ectoparasitic copepod, have become the major economic problem in the production of Atlantic salmon (*Salmo salar* L.) in Scotland, Ireland and Norway (Wootten *et al.*, 1982; Tully, 1989). The development of populations resistant to chemical treatments (Jones *et al.*, 1992), and increased environmental concern over the widespread use of pesticides, means there is a continued demand to develop new strategies in the control of sea lice. Although future developments may replace current chemical treatments, there will be a continuing need for cost-effective biological control in combined management programmes, particularly with the increasing trends towards producing salmon by chemical-free methods.

Observations of cleaning behaviour by British wrasse (Labridae) species were first reported in corkwing, *Crenilabrus melops* (L.), goldsinny, *Ctenolabrus rupestris* (L.) and rock cook *Centrolabrus exoletus* (L.) (Potts, 1973). Biological control of sea lice infestations in farmed salmon using native wrasse as cleaner-fish was first investigated in Norway (Bjordal, 1988, 1991). Since these early trials the use of wrasse, particularly the goldsinny, has been widely adopted on salmon farms throughout Scotland, Norway and Ireland (Bjordal & Costello, 1990; Darwall *et al.*, 1991; Treasurer, 1993). Successful sea trials have also been carried out using corkwing wrasse (Darwall *et al.*, 1991).

The demand for an adequate year-round supply of wrasse to the salmon farming industry may only be met by commercial breeding programmes as fisheries for wrasse are restricted to the summer months. Captive breeding would ensure continuous supplies and allow salmon cages to be stocked with wrasse with the introduction of new smolts in spring. In addition, fisheries for wrasse may not be locally sustainable and over-exploitation of stocks may give cause for concern.

One of the potential risks of stocking salmon cages with wild wrasse is the introduction of disease, particularly if they are caught from areas adjacent to salmon farms and transferred elsewhere. Although wild wrasse have been found to carry only atypical strains of *Aeromonas salmonicida*, which are not pathogenic to salmon (Frerichs *et al.*, 1992), wrasse removed from sea cages have been found with pathogenic strains of *A. salmonicida* derived from salmon (Treasurer & Cox, 1991; Treasurer & Laidler, 1994). As escapes of wrasse from salmon cages may be high, catching wrasse from the vicinity of a farm could result in the reintroduction of disease because escaped or wild fish may act as disease reservoirs following treatments or fallowing of salmon cages. Therefore, the risk of disease transmission could be greatly reduced by the introduction of captive-bred, certified disease-free stock.

The feasibility of breeding wrasse on a commercial scale was examined over a two-year period at a marine fish hatchery in Scotland (Golden Sea Produce, Hunterston, Ayrshire). Trials were carried out on goldsinny and corkwing wrasse using methods adapted from those already established for the commercial production of turbot [*Scopthalamus maximus* (L.)].

Although the goldsinny has been the most widely used species in the control of sea lice, rearing of corkwing wrasse was considered worthwhile because of their reported cleaning ability, larger size and hence potential for faster growth. Wild caught broodstock corkwing held at Hunterston continued to feed at lower temperatures than the goldsinny and were therefore thought to have greater potential to maintain cleaning activity during the winter. In addition, their larger size (approximately 12–18 cm in length) means they could be suitable for stocking with large, broodstock or market size salmon where problems have been experienced with predation of goldsinny (Treasurer, 1991a). This would also reduce escapement of wrasse or allow the use of larger mesh sizes than the 12–15 mm mesh necessary for goldsinny.

12.2   Materials and methods

12.2.1   *Broodstock*

Broodstock goldsinny of 9–14 cm and corkwing of 10–18 cm total length were caught from areas around the Firth of Clyde, Weymouth and Plymouth using fyke nets set at depths of 3–10 m. Stocks of both species were held in aquaria or tanks up to 3 m dia. under ambient light and temperature conditions to induce spawning in May and June as occurs in wild populations (Dipper & Pullin, 1979; Hilldén, 1981).

Stocks of goldsinny and corkwing were also held under three or two-month advanced photoperiod regimes so that the longest day-lengths occurred in March and April. Water temperatures were gradually increased to 13–14°C, one month before the onset of spawning was expected. A further group of broodstock goldsinny was held on a two-month delayed photoperiod, with the longest day occurring in August. However, there was no means to prevent an increase in the water temperature in this group and they were subjected to the same temperature regimes as the stocks under ambient photoperiod. Fish were established under these controlled photoperiod regimes 9–24 months prior to spawning.

Broodstock wrasse were maintained largely on rations of fresh or frozen ragworm, mussel or crabmeat supplemented with fresh herring roe when available. Broodstock were fed to satiation at rates ranging from approximately 5% of body weight per day in summer to <1% in mid-winter. Feeding was carried out on alternate days in the winter, increasing to twice daily in the spring and summer.

12.2.2   *Egg collection, fertilization and incubation*

Seawater used for egg collection, egg incubation and larval rearing was filtered to 5 μm and treated by ultraviolet (UV) sterilization. Eggs were collected from goldsinny and corkwing by manual stripping of gravid females; for convenience this is referred to as the spawning time.

Goldsinny eggs were stripped directly into glass beakers filled with seawater. The eggs were fertilized by stripping milt directly from males into the containers. These were gently mixed and left to stand for one hour. At the end of this period non-viable eggs settled at the bottom of the container while fertilized, viable eggs floated. The non-viable eggs were removed by siphoning and counted by subsample. The remaining viable eggs were counted by sub-sample as they rose to the surface of a test tube. The counts of viable and non-viable eggs were used to obtain the mean fecundity. The non-viable eggs were discarded while the fertilized viable eggs were transferred to 10 l plastic tanks filled with seawater. The tanks were kept in a constant temperature room at 15°C (± 1°C) and the water gently aerated. There was no water flow but the incubation tanks were siphoned daily to remove further non-viable eggs and 50–75% of the water was replaced daily.

Goldsinny larvae were later successfully hatched using a 20 l plastic tank with a flow-through system incorporating a 45 μm mesh on the outflow and a flow rate of 0.1 l min^{-1}. This improved hygiene and reduced the time involved in changing water.

Corkwing eggs were stripped directly onto 23 cm^2 plastic bioassay dishes filled with clean seawater. The eggs were fertilized by stripping milt directly from the males into the dishes. These were gently mixed and left to stand for 20–30 minutes to allow the eggs to adhere to the surface. The dishes were then rinsed in clean seawater and transferred to 20 l tanks. A gentle flow of water (0.15 l min^{-1}) from a recirculating filtration unit was passed over the eggs and a small airstone placed in front of the dishes to simulate the fanning action of the male corkwing in the nest (Potts, 1984). Eggs were incubated at around 15°C until hatching. Dead or infected eggs were removed daily by siphoning. Eggs could usually be stripped from a female so that small discrete clusters of eggs were spread over the dishes. Fecundity was estimated by counting the number of eggs in a single average sized cluster and multiplying the count by the number of clusters.

Once the goldsinny or corkwing larvae had hatched, the tanks were floated in the larval rearing tank to allow thermal equilibration before transferring the larvae directly into the main rearing tank.

To monitor the fecundity of individual goldsinny, two groups of fish (length range 10.6–12.4 cm, $n = 12$) were individually marked and held in separate aquaria under identical conditions and stocking densities. The first group was stripped each day while the second group was stripped every third day and the daily fecundity of each individual fish recorded.

12.2.3   *Larval rearing*

All ages quoted refer to days post-hatch, starting at day zero. Yolk-sac larvae were transferred on the day of hatching (day zero) into larval rearing tanks. Goldsinny larvae were reared in plastic 500 or 2000 l tanks with depths of 0.65 or 1.25 m respectively at densities of 1–110 larvae l^{-1}. Corkwing larvae were reared in 500 l tanks with a depth of 0.65 m at a density of 1–6 larvae l^{-1}. Larval rearing tanks were either static or had very low flow rates of 0.5 l min^{-1} (exchange rate 1.44 day^{-1}). All tanks were filled with filtered, UV-treated seawater of ambient salinity (32–4‰) at an initial temperature of 15°C and gradually increased to 19°C over a period of four to six days. Temperature, pH and dissolved oxygen were recorded daily and ammonia, nitrite and nitrate levels monitored as required.

Live feed organisms were obtained from large scale, bag cultures reared routinely for the commercial production of turbot. The larval tanks were stocked with the marine algae *Isochrysis galbana* Parke and maintained at densities of 60–120 × 10^6 cells per litre. From day zero, rotifers (*Brachionus plicatilis* Mueller), reared on the marine algae *Rhodomonas sp.* Karskeu, were introduced to the tanks and maintained, where possible, at 1–2 rotifers ml^{-1} for the first 30 days.

As an alternative first feed to rotifers, some goldsinny were reared in 2000 l or 20 000 l (6 m diameter) tanks stocked three to five weeks in advance with the copepod *Eurytemora velox* (Lilljeborg). These adult *Eurytemora* were derived from zooplankton trawls and maintained in 2000 l cultures stocked with marine algae. Copepod nauplii reached densities of $0.08-0.1 \, \text{ml}^{-1}$ by the time the goldsinny larvae were introduced.

Live *Artemia salina* L. were hatched from commercial dry cysts and fed either as nauplii after 18 hours incubation, or ongrown for a further 24 hours and enriched with a fish oil, soluble protein formulation. Adult *Artemia* were reared and enriched on the marine algae *Isochrysis* and *Rhodomonas*.

From day 20 in the corkwing tanks, and day 25 in the goldsinny tanks, *Artemia* nauplii were gradually introduced daily and, after a further two to three days, 24-hr ongrown *Artemia* were added twice daily. The introduction of rotifers and *Artemia* nauplii was then phased out over the next five to ten days.

Artificial shelters were suspended at the edge of the tanks, around day 20, constructed of blocks of plastic pipe of approximately 2 cm dia. and 6 cm in length. Where present, water flows were gradually increased to $3 \, \text{l} \, \text{min}^{-1}$ after day 25.

A trial was carried out to compare the survival and growth of goldsinny larvae in 20 l tanks stocked with the marine algae *I. galbana* and in control tanks without algae. Algae was introduced from day two onwards and larvae were sampled daily. Data for this trial were examined by a correlation test for normality (equivalent to the Shapiro-Wilk test) and an F-test for the homogeneity of variance. Statistical analysis of larval lengths was by the non-parametric Wilcoxon two-sample test.

12.2.4   *Ongrowing*

Once established on *Artemia*, the wrasse were referred to as juveniles. This was defined by a change in behaviour in both species when they started to settle around fixed objects, and also in the goldsinny by a change in appearance.

Goldsinny juveniles were transferred 36–58 days after hatching to clean 1 m or 2 m tanks at densities of up to $800 \, \text{m}^{-3}$. Corkwing juveniles were transferred 30–50 days after hatching to clean 1 m or 2 m tanks at densities up to $220 \, \text{m}^{-3}$. Temperatures were maintained at 19°C and flow rates were approximately $3-4 \, \text{l} \, \text{min}^{-1}$. Sections of plastic pipe 2–6 cm diameter were placed in the tank to provide shelter for the juveniles.

Goldsinny tanks continued to be stocked with ongrown *Artemia* until the juveniles were weaned on to non-live feeds. At this stage commercial dry crumb or 1.0 mm pellets were introduced and supplemented occasionally with fresh or frozen crab, mussel (*Mytilus edulis* L.) or ragworm [*Nereis diversicolor* (Müller)]. Corkwing tanks were stocked twice daily with ongrown *Artemia*, but this was gradually replaced with non-live feeds consisting largely of fresh or frozen crab, mussel or ragworm, supplemented occasionally with live adult *Artemia*.

Measurements of larvae were made on fresh unfixed material using an eyepiece graticule calibrated against a stage micrometer. The total length of larvae was measured from the tip of the lower mandible to the most posterior edge of the caudal fin. The mean weight of goldsinny juveniles was calculated by counting them into containers of seawater of known weight. The specific growth rate was calculated as follows:

$$SGR = \frac{\ln W_t - \ln W_0}{t} \times 100$$

where W_0 = initial weight, W_t = weight after t days, and t = number of days growth.

Mean lengths were determined by measuring individual fish held in a fine mesh net. Anaesthetics were not used for stripping of broodstock, nor length/weight determinations, nor for the transfer of fish between tanks. Most length/weight measurements of juvenile corkwings were obtained from mortalities.

12.3 Results

12.3.1 *Goldsinny egg production*

Small numbers of viable, fertilized goldsinny eggs were collected from natural spawning activity in tanks or purpose-built spawning aquaria. However, eggs collected in this way tended to accumulate a coating of debris and organic matter, irrespective of water quality. This contamination resulted in rapid deterioration of the eggs and none were successfully hatched. The majority of eggs produced by natural spawning appeared to be unfertilized, though spawning behaviour was frequently observed. Viable eggs were clear in appearance and buoyant, and most underwent normal cell division and hatching. Eggs which appeared cloudy at, or shortly after, stripping would sink to the bottom of the container. These were either unfertilized or non-viable eggs and did not undergo normal cell division and development.

All eggs used in trials were collected by manual stripping of broodstock fish. Eggs were usually collected early in the afternoon as this was when most spawning activity was observed. Eggs were first obtained from fish on a three-month advanced photoperiod and temperature regime on 18 February; from those on a two-month advanced regime on 16 March; and from those on ambient regimes on 7 May. Goldsinny held on a two-month delayed photoperiod but at ambient temperatures, spawned at the same time as fish on ambient photoperiod and temperature regimes. The fecundity of fish on advanced photoperiods was similar to that of fish held under ambient conditions, although fish held on a delayed photoperiod showed relatively poor and erratic egg production.

The largest egg yields were obtained from fish stripped within one to two days of capture, though egg production then abruptly halted in these stocks until the following year. The most consistent egg production was seen in fish which had been held in tanks over one to two years. Good egg production was clearly dependent on the health of the fish. A number of broodstock goldsinny which had deteriorated in condition were found to have an atypical strain of *A. salmonicida* (not pathogenic to salmon). These fish deteriorated rapidly in condition and egg production was very poor, with a mean fecundity of 1067 eggs per female (up to 49% viable) compared to a similar stock of non-infected fish with a mean fecundity of 3373 (up to 99% viable). Juvenile goldsinny reared at 19°C were stress tested for *A. salmonicida* by transferring them directly to water at 10°C and sampling the internal organs 10–14 days later. Juvenile goldsinny reared in the previous year, from broodstock fish later found to be infected with *A. salmonicida*, revealed no evidence of vertical transmission of the disease.

In the comparative trial to examine stripping of individual goldsinny, egg production commenced on 9 June and ended abruptly on 5 August. Goldsinny stripped each day or every third day, tolerated frequent stripping well, with only two male mortalities which were not related to stripping. All fish continued to feed well, and mean weight-loss in females was 4% in the first group and 3.6% in the second group. Daily stripping resulted in higher egg yields (mean fecundity 30 822) than in fish stripped every third day (mean fecundity 16 289). The fecundity of individual females ranged from 2700–81 000 eggs in one season with over 7000 eggs being collected in a single day from individual females. Some females produced good numbers of eggs throughout the season while others produced only one or two batches. The average daily proportion of females in a group producing eggs was 34%.

The percentage of viable eggs collected was similar in both groups, 57% when stripped daily and 50% when stripped every third day. Repeated stripping of males resulted in most individuals producing milt on every occasion, though the volumes obtained became very small as the season progressed.

The percentage of viable, fertilized eggs collected by manual stripping varied greatly between individuals but overall approximately 49% of all eggs collected in 1992 were estimated to be viable. Viable eggs were successfully fertilized up to 30 minutes after stripping (43% fertilized), though higher fertilization rates of 75% were achieved when eggs were fertilized within 10 minutes. Motility of sperm was reduced within 15 minutes of collection. Pooled batches of eggs were usually fertilized within five to ten minutes with freshly collected milt from three to four males.

Hatch rates of viable eggs ranged from 15 to 98% (mean 45%). Generally, eggs which failed to develop were evident within 24 hours and most of the eggs remaining after this time hatched provided no contamination occurred. Fertilized eggs had a mean (± SD) diameter of 0.85 ± 0.05 mm. Fig. 12.1 shows developing goldsinny eggs at 48 hrs post-fertilization. The first larvae hatched within 48 hrs (30° days) and most larvae had hatched within 72 hours at 15°C (45° days).

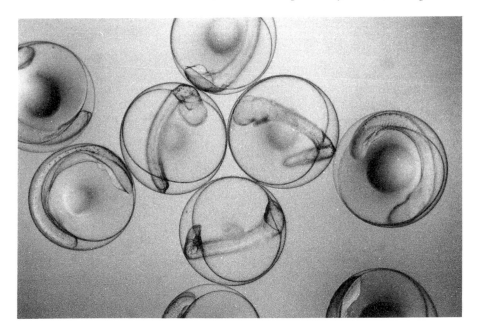

Fig. 12.1   Goldsinny eggs at 48 hours post-fertilization; actual size 0.85 mm dia.

12.3.2   *Goldsinny larval rearing and ongrowing*

Newly-hatched larvae at day zero were 1.8–2.3 mm in length (Fig. 12.2). The eyes, mouth and gut were poorly developed. The yolk sac was relatively large with a body depth of 0.8 mm. On day three the jaw was starting to develop and the gut appeared as a simple tube. Four patches of pigment became apparent; one above the gut, one above the anus and two in the caudal region. By day four the pectoral fins were visible, the yolk-sac was greatly reduced, the mouth open and algal cells were seen in the gut in approximately 30% of larvae (Fig. 12.3). At this stage the larvae were responsive and difficult to catch although they spent most of the time resting in a head down position. By day five the yolk-sac was fully absorbed and on day six the larvae developed dark vertical bands along the body.

It is uncertain whether algal cells were taken up passively or by active feeding as active uptake of algae was not observed. Survival of larvae in a tank stocked with *Isochrysis* and a control tank without algae was 96% in both groups at day five. On day six the survival rate was 77% in the tank with algae and 43% in the control tank. This difference in survival rates is comparable to the incidence of larvae with algal cells present in the gut. By day eight survival was reduced to 60% in the *Isochrysis* group and 43% in the control group (without *Isochrysis*). On day nine there were no surviving larvae in the control group and only 15% remaining in the fed group, and by day ten there were no surviving larvae in either group.

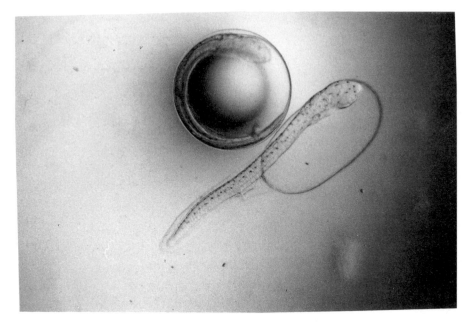

Fig. 12.2   Hatching goldsinny larvae (day zero); actual length 2.0 mm.

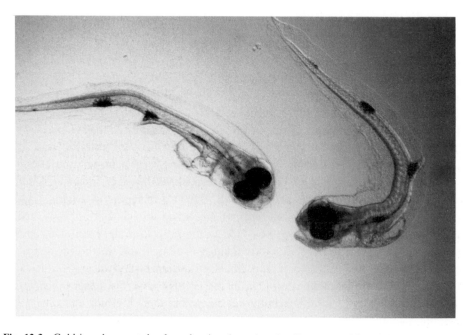

Fig. 12.3   Goldsinny larvae at day four showing the reduced yolk-sac; actual length 2.88 mm.

Larvae in both groups showed an increase in body length up to day five due to utilization of the yolk-sac reserves but larvae in the tanks with *Isochrysis* had a longer mean length by day four than those in the control tank (Table 12.1a). This coincides with the development of the mouth and the first observation of algal cells in the gut. However, in the tanks stocked with *Isochrysis*, algal cells were observed in only 5–30% of the larvae. The body length of the larvae with algal cells present in the gut, was greater in all cases than the mean length for that day. This would make the mean an unreliable indicator for the comparison of larval growth, and the maximum larval length may be a better indicator of growth. By days six and seven the maximum body lengths were 4% greater in the larvae in tanks with *Isochrysis* than in the control group and by day 8 the difference was 6% (3.25 mm compared to 3.05 mm). Despite the inclusion of data from larvae which did not appear to have fed, the mean lengths of goldsinny larvae in the tank with *Isochrysis* were significantly higher ($P < 0.05$) than in the control tank on days four, six and seven. However, by day eight the difference was no longer statistically significant (Table 12.1b). Rotifers were first found in the gut on day seven. At this stage, rotifers were clearly a very large prey item for goldsinny larvae as a single rotifer virtually filled the gut (Fig. 12.4), though by day twelve many rotifers were visible in single guts. Availability of rotifers of a suitable size was considered to be the most significant factor in larval survival at this stage. Larvae did not appear to survive when rotifers were introduced later than nine days post-hatch.

When goldsinny larvae were first transferred to the rearing tanks they remained visible in large numbers near the surface of the water. Generally larvae were not seen between days four to twenty and in one case this 'disappearance' phase lasted up to day 43. Only when there was a large variation in the ages of the larvae did some remain visible throughout. This necessitated the maintenance of tanks and provision of live-feeds over a long period of time when the survival of larvae was unknown. The highest rate of larval mortality occurred during this stage and because of the rapid decomposition of the larvae it was difficult to determine cause of mortality. The post-larvae generally reappeared around day 20 (Fig. 12.5) and shortly afterwards (around day 25) underwent a further change in appearance becoming dark brown in colour with a horizontal gold stripe extending from the eye along the length of the body. This was accompanied by a change in behaviour with the goldsinny moving to the sides of the tanks and resting on or around any available objects in the tanks which provided shelter. At this stage the goldsinny were approximately 10 mm in length and were referred to as juveniles. The age at which this metamorphosis occurred was very variable between 25 and 45 days but in a single tank all larvae would change in appearance within three to four days despite age differences of up to 19 days. This change in behaviour was a good indicator to commence feeding on *Artemia* nauplii. Occasionally some of the post-larvae which had not yet metamorphosed were observed to have fed on *Artemia* nauplii without any apparent problems, and the introduction of *Artemia*

Table 12.1a Mean length, standard deviation and ranges of goldsinny larvae with and without the marine algae *Isochrysis galbana*.

Group	Length (mm)	Day 0	Day 1	Day 2	Day 3	Day 4	Day 5	Day 6	Day 7	Day 8
With algae[1]	Mean	2.35	2.40	2.42	2.55	2.67	2.60	2.81	2.82	2.79
	Std. Dev.	0.03	0.03	0.05	0.16	0.13	0.21	0.22	0.16	0.25
	Min.	2.31	2.36	2.37	2.30	2.59	2.32	2.47	2.62	2.37
	Max.	2.38	2.44	2.50	2.90	2.90	3.00	3.10	3.10	3.25
	n	7	8	10	10	10	10	19	10	14
No algae	Mean	2.35	2.40	2.42	2.55	2.52	2.65	2.52	2.51	2.63
	Std. Dev.	0.03	0.03	0.06	0.23	0.18	0.17	0.22	0.22	0.24
	Min.	2.31	2.31	2.35	2.10	2.25	2.45	2.12	2.05	2.27
	Max.	2.38	2.36	2.55	2.87	2.85	2.97	2.97	2.97	3.05
	n	7	8	10	10	12	10	21	14	15

[1] Algae added from day two onwards.

Table 12.1b Results of Wilcoxon two-sample test on goldsinny length with and without the marine algae *Isochrysis galbana*.

Group	Day 0	Day 1	Day 2	Day 3	Day 4	Day 5	Day 6	Day 7	Day 8
Algae vs. No Algae	$U_S = 24.5$ $P > 0.05$	$U_S = 32.0$ $P > 0.05$	$U_S = 53.0$ $P > 0.05$	$U_S = 53.0$ $P > 0.05$	$U_S = 93.0$ $P < 0.05$	$U_S = 54.5$ $P > 0.05$	$U_S = 320.0$ $P < 0.002$	$U_S = 124.0$ $P < 0.002$	$U_S = 143$ $P > 0.05$
Critical values	$U_{0.05(7,7)} = 41$	$U_{0.05(8,8)} = 51$	$U_{0.05(10,10)} = 77$	$U_{0.05(10,10)} = 77$	$U_{0.05(12,10)} = 91$	$U_{0.05(10,10)} = 77$	$U_{0.002(20,19)} = 312$	$U_{0.002(14,10)} = 121$	$U_{0.05(15,14)} = 151$

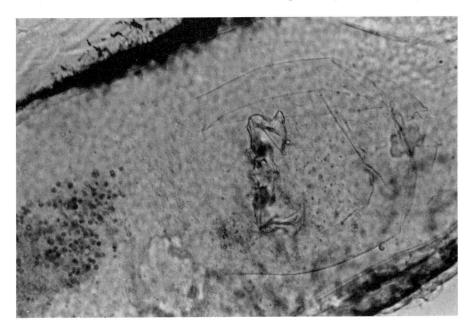

Fig. 12.4 Partially digested rotifer and algal cells in the gut of day seven goldsinny larvae. Rotifer length 300 μm. .

Fig. 12.5 Goldsinny post-larvae shortly before metamorphosis (day 20); actual length 8.5 mm.

may have been possible as early as day 20. Ongrown *Artemia* were introduced shortly afterwards and, after a further 10 to 20 days, the juveniles were transferred to ongrowing tanks. Survival of goldsinny from hatching to this post-artemia stage ranged from 0.1–7%.

Larvae reared on copepods, as an alternative feed to rotifers, did not appear to have a higher survival rate at first feeding. However, direct comparisons could not be made as densities of copepod nauplii were rarely greater than $0.1\,ml^{-1}$ compared to rotifer densities maintained at $1.0\,ml^{-1}$. Despite this low prey-density, relatively good survival (1.5–2.0%) was achieved. The larvae reared on copepods appeared to be more healthy and robust and had higher growth rates, with earlier metamorphosis (day 25), than those reared exclusively on rotifers. In one comparative trial, goldsinny reared exclusively on copepods had a mean weight of $0.44\,g$ on day 43 compared to those reared on rotifers which had a mean weight of $0.33\,g$ on day 43, despite being fed also on *Artemia* from day 29. Goldsinny juveniles reared on copepods had a distinct brick-red colouration.

Survival of goldsinny juveniles from the stage of metamorphosis at day 40 to day 150 was generally around 95% and once the larvae were feeding on *Artemia* few problems were experienced. However, in the first year of trials problems were experienced in some groups with high mortalities occurring following the introduction of *Artemia*. The post-larvae became very delicate and died instantly following any minor disturbances. This was attributed to a nutritional problem associated with a single batch of *Artemia* cysts and mortalities stopped within three to four days of introducing a different strain of *Artemia*. Of 13 480 larvae which reached metamorphosis in these tanks (overall 3.6% survival from hatching) only 2500 survived to the post *Artemia* stage (0.9% survival from hatching). No similar problems were encountered in the second year following the introduction of *Artemia*.

Once established on *Artemia* the juveniles were transferred to ongrowing tanks and weaned onto commercial dry pellet feeds. Only occasional mortalities occurred in this stage largely due to ciliate infections or from fish jumping out of the tanks. The ciliate problem was overcome by providing shelters to discourage the juveniles from lying in waste debris on the bottom of the tanks and by occasional formalin treatments at a dose rate of 200 ppm for 20 minutes. Goldsinny tended to jump out of the water particularly if disturbed at night, and goldsinny of only 4–5 cm frequently jumped out of tanks with a clearance of around 20–30 cm. If the water level could not be kept low enough, the tanks were covered and disturbance of the fish avoided in poor light conditions. Although confrontational behaviour did occur, maintenance of goldsinny juveniles at densities of up to $800\,m^{-3}$ did not result in any apparent injuries or mortalities and cannibalism was not observed.

Figure 12.6 shows the weight and length data obtained from these trials for cultured goldsinny. Pooled data for age and mean length of cultured goldsinny are shown in Fig. 12.7. Goldsinny juveniles, reared at 19°C, reached a mean (±SD)

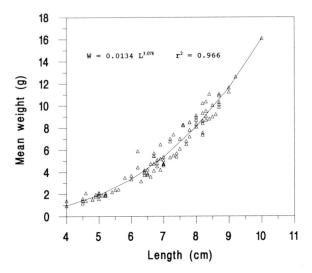

Fig. 12.6   Weight and length data for cultured goldsinny wrasse.

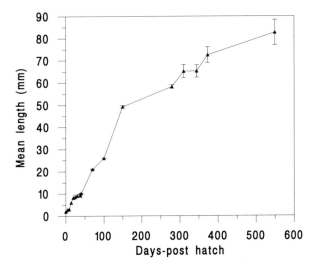

Fig. 12.7   Mean (±SD) length and age data for goldsinny reared at 19°C.

length of 8.26 cm ± 0.40 by 540 days post-hatch although some individuals were over 9 cm.

The growth of hatchery reared goldsinny juveniles held at ambient temperatures was reduced overwinter, when temperatures were as low as 6°C and may have reached 4°C overnight. The overall specific growth rate (SGR) at ambient temperatures was 0.6% body weight per day. There was virtually no growth over the winter and the SGR was only 0.1% between December and April. Despite

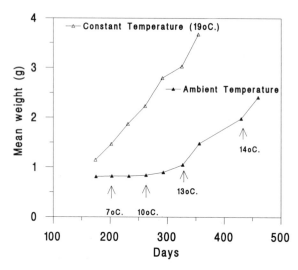

Fig. 12.8 Mean pooled weights of goldsinny reared under ambient or constant (19°C) temperatures.

poor food availability the SGR increased to 1.0% between May and September, when the mean weight increased from 1.05 g to 2.41 g as temperatures rose to 12–14°C (Fig. 12.8). In comparison goldsinny juveniles of approximately the same size and age reared at 19°C had a specific growth rate of 1.4% over the same period. Specific growth rates for goldsinny at 19°C increased from 1.3% (days 94–216) to 2.0% (days 270–550) as the juveniles became larger.

The best growth rates of goldsinny were achieved using a commercial dry pellet feed (SGR 1.32%) compared to a diet of dried amphipod meal (SGR 0.93%) or a diet of fresh mixed shellfish and ragworm (SGR 0.89%). Although the amphipod meal was readily accepted by the goldsinny it proved difficult to consume because of its high buoyancy. This required a learning period in which the juveniles had to develop a feeding technique and this may have contributed to the lower SGR. In addition, a large number of juveniles failed to adapt to the amphipod meal, became emaciated and had to be removed from the trial.

12.3.3 *Corkwing egg production*

No attempt was made to manipulate the ratio of sexes in broodstock tanks and this was solely dependant on availability. Generally the ratio of males to females was around 1:4, but in one tank the ratio was 2:1. Although territorial behaviour was observed in male corkwings no severe lesions or mortalities were attributed to aggression.

Examination of the fish showed approximately 10% of male corkwings to be satellite or type-2 males, that is functional males with a female appearance and abdominal swelling during the spawning season (Dipper & Pullin, 1979). Most of these satellite males produced very large volumes of viable milt (up to 2 ml from one individual on a single day). These type-2 males were all smaller than the normal type-1 males, and showed marked abdominal distension during the spawning season. They could only be distinguished from females by the large volumes of milt produced during manual stripping. Nest-building behaviour by male corkwing wrasse was observed early in the spawning season. Nests were constructed in lengths of 6″ diameter pipes with *Fucus sp.* seaweeds provided, though eggs were never found in these nests as spawning may have been disrupted by removal of the pipes for examination and the capture of fish for manual stripping. Corkwing wrasse adapted poorly to glass aquaria and nest-building did not occur under these conditions.

All eggs used in the trials were collected by manual stripping of broodstock fish. Eggs were first obtained in the first week of March in the three-month advanced photoperiod stocks, in mid-April in the two-month advanced stocks, and in the first week of June in the corkwings held under ambient conditions. Fish were stripped at two to four day intervals, depending on appearance, with an average of 48% of females producing eggs on each stripping. Most females gave two to three large batches of up to 1000 eggs, although some females yielded several smaller batches of eggs throughout the spawning period.

The overall mean fecundity in corkwings acclimated in tanks for two years and held at a stocking density of 1.4 females per m^{-3} was approximately 3200 eggs per female, with an average of 71% of females yielding eggs when stripped. In comparison, corkwings acclimated for only one year and held at a higher stocking density of 12.8 females m^{-3}, had a mean fecundity of 400 eggs per female with only 25% of females yielding eggs. Both of these groups were held under ambient photoperiod and temperature conditions.

Corkwings held under three-month advanced photoperiod and temperature regimes had a mean fecundity of approximately 1200 eggs per female with an average of 48% of females yielding eggs. These fish were also held for two years at a stocking density of 1.4 females m^{-3}. No data was collected on the corkwings held under two-month advanced photoperiod regimes.

In one year, a total of 60 000 eggs was obtained from 78 female corkwings. Unfortunately, a power failure in the incubation system resulted in the loss of many of these eggs and only 18 000 survived to hatching. Only 4% of eggs were hatched successfully in the first year but in the second year, with an improved incubation system, an overall hatch rate of 51% was achieved with individual batches ranging from 7% to 96%.

Corkwing wrasse produce sticky demersal eggs with a diameter of 0.9 mm at fertilization, increasing to 1.0 mm just prior to hatching (Fig. 12.9). Incubation times ranged from 9–15 days at 14–15°C although hatching times within a single days batch (multiple females) varied by two to six days.

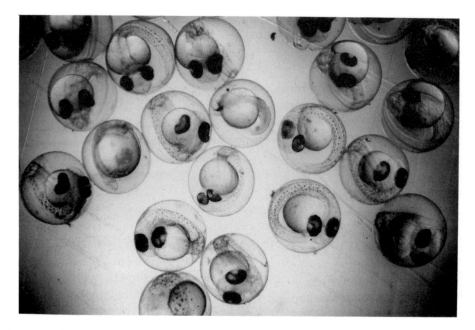

Fig. 12.9   Corkwing eggs at eight days post-fertilization; actual size 0.9 mm dia.

12.3.4   *Corkwing larval rearing and ongrowing*

Corkwing larvae at hatching were fairly robust and could be transferred to the larval tanks in small beakers or by picking them out with a wide bore siphon using a very slow flow rate. This second method also made it possible to accurately count the larvae. Larvae which were not transferred to the larval rearing tanks on the day of hatching showed very poor survival.

The hatched larvae were larger (2.6–2.8 mm in length) and more developed than the goldsinny larvae. The yolk-sac was small with a body depth at this point of 0.4 mm. The mouth was well developed and the eyes appeared to be functional. By the second day the yolk-sac was no longer present and the pectoral fins were visible (Fig. 12.10). The presence of algae in the gut was first noticed on day three following hatching, and partially digested rotifers were observed on day five. By day eight the larvae had doubled in size since hatching and the anal, dorsal and pectoral fins were developing (Fig. 12.11).

Early growth of corkwing larvae was rapid and artemia nauplii were introduced as early as day 18. Survival of corkwing larvae from hatching to around day 40 was relatively high ranging from 7–32%. The larvae remained visible in the tanks throughout the rearing period.

Around days 30–35 when they were 10–12 mm in length the larvae underwent behaviourial changes, although unlike goldsinny they showed little change in morphological appearance. The larvae moved from the open water to the sides of

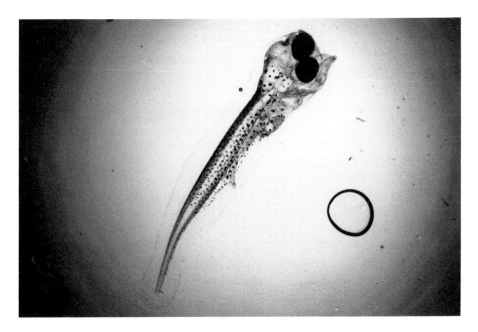

Fig. 12.10   Corkwing larvae (day two); actual length 2.9 mm.

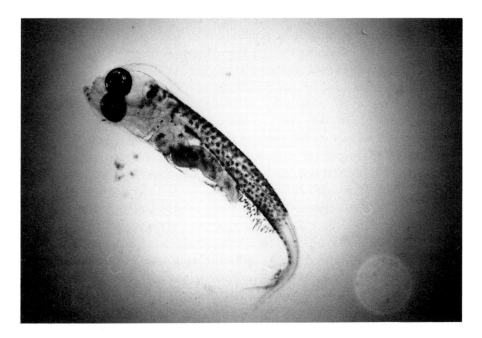

Fig. 12.11   Corkwing larvae (day eight); actual length 5.2 mm.

the tanks or settled around any fixed objects and at this stage could be transferred to clean ongrowing tanks.

The sections of plastic pipes were rapidly adopted by the juvenile corkwings and this discouraged them from lying on the bottom of the tanks which accumulated organic debris and could be a source of ciliate or nematode infestation. After 10–20 days on *Artemia* the juvenile corkwings were transferred to clean ongrowing tanks.

In the first year of trials high mortalities occurred following transfer from the larval rearing tanks to ongrowing tanks. While mortality rates in goldsinny larvae were only high during the early first feeding phase, 50% of mortalities in corkwing occurred at a later stage, between days 40–76. This resulted in an overall survival of 14% in the first year with 120 larvae out of 816 hatched surviving to 76 days post hatching. Initially these mortalities were attributed to ciliate and nematode infections but these were only found occasionally in moribund fish compared to the high numbers found in dead fish. Mortalities were reduced in the second year when large, deeper tanks became available for the ongrowing phase. These tanks were located in an area less prone to disturbance and light was largely excluded by dark tank covers. Tank cleaning and disturbance was kept to a minimum. These reductions in stress greatly improved survival and growth, with only a 9% mortality rate occurring between approximately days 40–150 and a best survival achieved of 10% from hatching to 150 days. Corkwing juveniles in this group were ongrown for a period with a large number of goldsinny juveniles. These were slightly smaller than the corkwings, and though some confrontations were observed, particularly during feeding, these did not result in any apparent injuries or mortalities. Generally, the smaller goldsinny usually came off best in disagreements over food possession, though the corkwings were more successful at defending their adopted hiding places. There was no evidence of cannibalism in corkwing at the juvenile stage, even when the ages in a single tank differed by 19 days.

In the second year of trials, the best survival achieved was 32% from hatching to day 41. There was an overall survival from 7800 hatched larvae, to day 40 of 1010 juveniles (13%) and 407 of these survived to a mean age of 150 days (5% survival). The overall survival from 18 000 eggs to approximately 150 days was 2.3%.

Survival would have been much higher in the second year if the largest batch of corkwing juveniles had not been moved to ongrowing tanks too early, owing to limited availability of tanks, resulting in high mortalities in this group. In addition the larger ongrowing tanks were not available until later in the season and the improved late survival by reducing stress had not yet been identified.

Shortly after transfer the juveniles accepted non-live feeds of chopped mussel, crab or amphipods, supplemented occasionally with live adult *Artemia*. They did not readily accept fish pellets in dry or semi-moist form.

As efforts were largely concentrated on improving the survival of corkwing very little data was collected on growth rates in order to minimize stress. Although many of the measurements of older fish were based on mortalities, those which showed poor body condition were excluded. Initial findings suggest that artificially reared corkwings may reach lengths of over 10 cm within 12 months (Fig. 12.12).

Table 12.2 summarises the development and culture of goldsinny and corkwing wrasse.

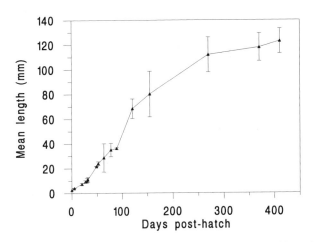

Fig. 12.12   Mean (±SD) length-at-age data for corkwing wrasse reared in 1991 and 1992.

Table 12.2   Summary of the development and culture of goldsinny and corkwing wrasse.

	Goldsinny	Corkwing
Spawning	mid-May–late July	early–late June
Season	Continuous	batch spawners?
Fecundity	3000–80 000	500–3000
% eggs viable	50	90
Egg type	pelagic, buoyant	demersal, adherent
Diameter (mm)	0.85	0.90
Hatching		
Length at day zero (mm)	2.0	2.6
Yolk-sac	large	small
Yolk absorbed	day 5	day 2
Algae in gut	day 4	day 3
Rotifers in gut	day 7	day 5
First *Artemia* feed	day 25	day 20
Survival for days 0–40	0.1–7.0	7–32
Survival for days 40–150	95	50–90
First dry pellet-feed	day 40	not accepted
Predicted 10 cm	within 2 years	within 1 year

12.4   Discussion

12.4.1   *Goldsinny*

The small number of fertilized eggs collected from goldsinny broodstock tanks suggests the proportion of eggs fertilized during natural spawning is very low and this, coupled with the tendency for eggs to become contaminated, suggests collection of eggs by this method may be unsuitable. Hilldén (1981) observed that a majority of naturally spawned goldsinny eggs sink after spawning with only 10% buoyant. He suggested goldsinny produce both buoyant pelagic eggs and sinking demersal eggs which have the advantage of remaining in the territory. However, in this study, only clear eggs which remained buoyant underwent normal cell division and were hatched successfully, while eggs which were cloudy in appearance and sank were either non-viable or unfertilized.

Manipulation of photoperiod and temperature regimes permitted production of goldsinny eggs outside the natural spawning season. Manual stripping of goldsinny resulted in individual females producing eggs more or less continuously throughout the season. The failure of goldsinny, held on a two-month delayed photoperiod, to show a late spawning in August suggests temperature plays an important role in the regulation of spawning. The relatively poor and erratic egg production which occurred in this group was probably a result of the conflicting effects of photoperiod and the unavoidable increase in temperature.

The higher survival and longer larval lengths observed in tanks stocked with *Isochrysis*, compared to those without any feed source, suggests feeding on algal cells may be important in the growth and survival of goldsinny larvae in the first few days. However, this effect is probably replaced when feeding on rotifers commences. The failure of larvae to survive in the experimental group with *Isochrysis* beyond day nine may be because of a requirement at this stage for a larger prey such as rotifers, lack of environmental stability, accumulation of bacterial or organic contaminants, or poor survival in small experimental tanks. Larger tanks were unsuitable for continuous monitoring due to the difficulties of locating and catching live larvae. The low incidence of larvae with algal cells present in the gut suggests that many of the larvae in the tanks with *Isochrysis* failed to feed. This may have resulted in these individuals having a lower mean length than larvae which had fed on algae. The apparent absence of algal cells in the gut of many individuals suggests algae are consumed by an active feeding process rather than by passive uptake through drinking. Low larval survival or a later onset of starvation (because of possible dietary inadequacies of algae as a first feed) may explain the loss of a significant difference in growth rates of larvae at day eight.

The 'disappearance' of goldsinny larvae may be attributed to them maintaining a vertical position in the water at a depth where light just penetrated. In addition, the development of dark vertical bands around day six disrupted the visible outline of larvae and made them difficult to see in tanks stocked with algae. These may be adaptations associated with predator avoidance. The main factor in determining survival during this disappearance phase was considered to be a failure of first feeding related either to insufficient prey density or inappropriate size of feed organisms. It was uncertain whether the even metamorphosis of post-larvae was a result of synchronization of larval development or lower survival of younger larvae.

Possible improvements could be made to the early survival of goldsinny larvae during the first feeding phase by examining the use of alternative prey organisms and higher prey densities. Copepods would appear to be an ideal first-feed as their different stages make a range of prey sizes available to the developing larvae with the early nauplii stages only 130 μm in length compared to rotifers at 200–300 μm. It is difficult to make a direct comparison of survival rates because of the different densities of rotifers and copepod nauplii achieved. Relatively good survival and growth rates were achieved in tanks stocked only with adult copepods several weeks prior to the introduction of larvae, despite the very low densities of copepod nauplii present. Although reliable cultures of copepods are difficult to maintain, future developments using large scale mesocosms may represent a low technology, cost effective means of rearing wrasse and other species.

The mean length of 8.3 cm (both sexes) at 540 days was comparable to that of wild goldsinny from the same area. Male goldsinny, caught in the Millport area in the Firth of Clyde, had mean lengths of 8.1 cm at one year and 9.1 cm at two years, while females had mean lengths of 6.3 cm at one year and 7.7 cm at two years (Sayer *et al.*, 1995). Growth of goldsinny was probably enhanced by the high temperatures used and could possibly be further increased by even higher temperatures. Goldsinny reared at 19°C but accidentally exposed to temperatures of 24°C showed no adverse affects and had an increased feeding response. Rearing of goldsinny at these temperatures may necessitate an acclimation period before transfer to sea cages and exposure to ambient temperature. However, cultured goldsinny were found to be very resilient to sudden temperature changes. Direct transfer of goldsinny, reared at 19°C, to water at 10°C caused them to become stressed and lethargic but within two days normal behaviour and feeding was resumed, though at a reduced level. Turbot juveniles treated in the same way did not resume feeding and high mortalities occurred within a few days.

A minimum size for goldsinny wrasse of 10 cm has been suggested for stocking in salmon cages as only goldsinny greater than 8.9 cm in length are retained by a 12 mm mesh, with escapes or gilling of smaller fish occurring (Treasurer, 1991a).

It can be estimated from the available growth data that goldsinny reared at 19°C could reach a minimum mean length for stocking within two years. While improvements in rearing techniques or the use of higher temperatures may improve growth rates, selection of broodstock may also influence the mean length at age. Goldsinny in Swedish waters attained a length of 11.3 cm at two years of age (Hilldén, 1978) compared with a mean length of only 10.7 cm at two years in

Scotland (Treasurer, 1992). Although adult goldsinny are often 13–15 cm in length (Hilldén, 1978; Treasurer, 1992; Sayer *et al.*, 1995) most of the broodstock captured in the Clyde area for this study were less than 12 cm in length.

The slower growth rates in this area were confirmed by studies on the west coast of Scotland which compared goldsinny from Millport (adjacent to the broodstock catch area) with those from Oban and Luing in the north (Sayer *et al.*, 1995). Two-year-old goldsinny from the Millport area had mean lengths of 7.7 cm (females) and 9.1 cm (males) compared to mean lengths of 9.3 cm (females) and 9.6 cm (males) in the Oban area (Sayer *et al.*, 1995). These lower growth rates were thought to be a result of a higher level of investment in gonadal development. Goldsinny in Swedish waters also had a mean length of 9.5 cm at two years of age (Hilldén, 1978). Stocks of goldsinny from different areas may therefore show considerable variation in size although the size structure of fish obtained may also have been influenced by capture techniques. Goldsinny reared from broodstock fish collected in the Clyde area may have had a lower growth potential. Improved growth rates might be achieved in future by selection of broodstock fish from areas with faster growth rates.

12.4.2   Corkwing

Corkwing wrasse appear to be batch spawners but this impression may be due to the timing of manual stripping and needs further investigation. The culture of naturally spawned eggs appears to be unsuitable due to the difficulties in determining whether spawning has taken place without disturbing the fish. In addition, the tendency of corkwing wrasse to lay their eggs in seaweed would make hygienic incubation of eggs difficult.

The characteristics of type-2 males were described by Dipper and Pullin (1979) who recorded around 20% of corkwing males as type-2 males. In this study only 10% of males were type-2 or satellite males. The abdominal distension in these individuals is a result of the very large testes (Dipper & Pullin, 1979) and was confirmed in this study by the production of larger volumes of milt than in 'normal' type-1 males.

The long incubation period for corkwing eggs made them susceptible to contamination, although this was easily controlled by the removal of affected eggs. The development of a new incubation system in the second year greatly reduced the incidence of infection and improved egg survival and hatching rates. Research could be conducted on disinfection techniques for corkwing eggs to further reduce the incidence of infection.

The relatively high early survival of corkwing larvae, compared to goldsinny larvae, may be related to the greater degree of development at hatching and the suitability of rotifers as a first-feed organism, rather than to better artificial rearing techniques. Rotifers appear to be an appropriate first-feed organism for corkwing wrasse because of the large size and well-developed mouth of the larvae. The tendency of post-larvae, at around days 30–35, to move from the open water and settle around fixed objects probably relates to a change from a pelagic to a demersal habitat.

Experience has shown that corkwing wrasse are more timid and less resilient to stress than goldsinny wrasse. This can be overcome by achieving optimum condition of the fish, minimising disturbance and providing adequate shelter. After a critical period corkwing appear to become more tolerant to disturbance and this may give them a further advantage over wild caught wrasse in adapting to salmon farm operations.

The growth rate of corkwing was slightly higher than in wild corkwing juveniles (this volume, Chapter 2). The high growth rate allowed a size of 10 cm to be reached within one year. This is the minimum size recommended for stocking of goldsinny in cages with a 12 mm mesh (Treasurer, 1991b). The larger size of the corkwing may make it a more suitable species for stocking with large salmon or in cages where a larger mesh size is required. A minimum wrasse size of 12 cm is recommended for stocking with a 15 mm mesh (Darwall *et al.*, 1991). However as corkwing wrasse have a deeper body shape it may be possible to reduce the minimum stocking length for this species.

Although problems were initially encountered with rearing corkwing, particularly during the weaning phase, these were gradually resolved and good survival and growth rates were achieved by the end of the second season. This was largely a result of improvements in egg incubation and ongrowing and a better understanding of corkwing behaviour. It is predicted that survival rates of around 30% or more from hatching could be achieved in commercial rearing. Further improvements in survival and growth may also be achieved by improvements in live feed availability and by developing an acceptable artificial diet for the juveniles.

12.4.3   *Final conclusions*

In two years approximately 5000 goldsinny and 1000 corkwing juveniles were reared on an experimental basis. This suggests commercial rearing of these two wrasse species would be viable based on the available technology for live-feed production.

Although territorial behaviour and confrontations were frequently observed, the absence of serious aggression in both broodstock and juvenile wrasse may have been because the high stocking densities prevented the establishment of proper territories. This has the advantage of allowing high densities of fish to be held in relatively small tanks. However, the effects of mixing larvae of different ages at the first feeding stage is not known but cannibalism was considered unlikely as a very small prey size was required at this age. The absence of cannibalism in cultured wrasse is important in reducing the mortality, stress and disease which may be associated with cannibalism of juvenile goldsinny in the production of other commercial species such as turbot.

The high survival of corkwing larvae and the potential to reach a size suitable for stocking with salmon in one year suggests corkwing may be the more economically viable species for commercial rearing. At present, goldsinny are used more widely in salmon farming than any other species (Darwall *et al.*, 1991; Sayer *et al.*, 1993). Improved availability of corkwing wrasse through commercial breeding could change this situation particularly if corkwing wrasse prove to be a more appropriate species for stocking with larger salmon and for overwintering.

Acknowledgements

I would like to thank Dr James Bron, Dr Niall Bromage, Dr William Roy, Ronnie Soutar and, especially, the staff of Golden Sea Produce at Hunterston. Stock cultures of marine algae and rotifers were supplied by the Scottish Marine Biological Association, Oban. This project was jointly funded by Golden Sea Produce Ltd, the Scottish Salmon Growers Association and the Norwegian Fisheries Research Council.

References

Bjordal Å. (1988) Cleaning symbiosis between wrasse (Labridae) and lice infested salmon (*Salmo salar*) in mariculture. *International Council for the Exploration of the Sea, Mariculture Committee* 1988/F **17**: 8 pp.

Bjordal, Å. (1991) Wrasse as cleaner fish for farmed salmon. *Progress in Underwater Science* **16**: 17–28.

Bjordal, Å. and Costello, M.J. (1990) How good is this natural control of sea-lice? *Fish Farmer* **13(3)**: 44–6.

Darwall, W., Costello, M.J. and Lysaght, S. (1991) Wrasse–How well do they work? *Aquaculture Ireland* **5**: 26–9.

Dipper, F.A. and Pullin, R.S.V. (1979) Gonochorism and sex inversion in British Labridae (Pisces). *Journal of Zoology (London)* **187**: 97–112.

Frerichs, G.N., Millar, S.D. and McManus, C. (1992) Atypical *Aeromonas salmonicida* isolated from healthy wrasse (*Ctenolabrus rupestris*). *Bulletin of the European Association of Fish Pathologists* **12**: 48–9.

Hilldén, N.O. (1978) An age-length key for *Ctenolabrus rupestris* (L.) in Swedish waters. *Journal du Conseil international pour l'Exploration de la Mer* **38**: 270–71.

Hilldén, N.O. (1981) Territoriality and reproductive behaviour in the goldsinny *Ctenolabrus rupestris* L. *Behaviourial Processes* **6**: 207–21.

Jones, M.W., Sommerville, C. and Wootten, R. (1992) Reduced sensitivity of the salmon louse (*Lepeophtheirus salmonis*) to the organophosphate dichlorvos. *Journal of Fish Diseases* **15**: 197–202.

Potts, G.W. (1973) Cleaning symbiosis among British fish with special reference to *Crenilabrus melops* (Labridae). *Journal of the Marine Biological Association of the United Kingdom* **53**: 1–10.

Potts, G.W. (1984) Parental behaviour in temperate marine teleosts with special reference to the development of nest structures. In *Fish Reproduction, Strategies and Tactics* (Ed. by Potts, G.W. and Wootton, R.J.). Academic Press, London, pp. 223–44.

Sayer, M.D.J., Gibson, R.N. and Atkinson, R.J.A. (1993) Distribution and density of populations of goldsinny wrasse (*Ctenolabrus rupestris*) on the west coast of Scotland. *Journal of Fish Biology* **43** (Supplement A): 157–167.

Sayer, M.D.J., Gibson, R.N. and Atkinson, R.J.A. (1995) Growth, diet and condition of goldsinny *Ctenolabrus rupestris* (L.), on the west coast of Scotland. *Journal of Fish Biology* **46**: 317–40.

Treasurer, J. (1991a) Wrasse need due care and attention. *Fish Farmer* **14(4)**: 24–6.

Treasurer, J. (1991b) Limitations in the use of wrasse. *Fish Farmer* **14(5)**: 12–13.

Treasurer, J. (1993) More facts on the role of wrasse in louse control. *Fish Farmer* **16(1)**: 37–8.

Treasurer, J. and Cox, D. (1991) The occurrence of *Aeromonas salmonicida* in wrasse (Labridae) and implications for Atlantic salmon farming. *Bulletin of the European Association of Fish Pathologists* **11**: 208–10.

Treasurer, J. and Henderson, G. (1992) Wrasse: A new fishery. *World Fishing* **41**: 2.

Treasurer, J. and Laidler, L.A. (1994) *Aeromonas salmonicida* infection in wrasse (Labridae), used as cleaner fish, on an Atlantic salmon, *Salmo salar*, farm. *Journal of Fish Diseases* **17**: 155–61.

Tully, O. (1989) The succession of generations and growth of the caligid copepods *Lepeophtheirus salmonis* and *Caligus elongatus* parasitising farmed Atlantic salmon smolts (*Salmo salar* L.). *Journal of the Marine Biological Association of the United Kingdom* **69**: 601–12.

Wootten, R., Smith, J.W. and Needham, E.A. (1982) Aspects of the biology of the parasitic copepods *Lepeophtheirus salmonis* and *Caligus elongatus* on farmed salmonids and their treatment. *Proceedings of the Royal Society Edinburgh* **81B**: 185–97.

PART II
AQUACULTURE APPLICATIONS

Chapter 13
Development and future of cleaner-fish technology and other biological control techniques in fish farming

M.J. COSTELLO *Environmental Sciences Unit, Trinity College, Dublin 2, Ireland*

The ability of locally collected wrasse to clean sea lice infesting Atlantic salmon, *Salmo salar*, in fish farms was first demonstrated in Norway by Bjordal. This cleaner-fish technology was rapidly adopted by salmon farmers, such that in 1994 over 130 farm sites in Norway were using 1.5 million wrasse, and 30 farm sites in Scotland using 150000 wrasse.

With the development of intensive finfish culture worldwide, these and other species of cleaner-fish may prove valuable in controlling sea lice, other ectoparasites, and microbial skin diseases. Tilapia are one of the most widely cultured fish in tropical waters, and preliminary studies have shown tropical cleaner-fish to clean marine cultured tilapia. Many freshwater fish show cleaning behaviour, including carp, *Cyprinus carpio*, which is an important farm species. Wrasse in use as cleaner-fish can significantly reduce fouling on salmon cage nets. However, biological control has been in use to control fouling in shellfish culture before the discovery of cleaner-fish technology. These methods of biological control of parasites and fouling are natural, cheap, environmentally benign, utilize locally available resources, and can be easily adopted by fish farmers.

The future development of cleaner-fish technology is primarily limited by the availability of wrasse, and also by concerns over possible transfer of pathogens from wrasse to salmon. A source of cultured, certified disease-free wrasse, and more studies on wrasse parasites and diseases, may overcome these problems. Wrasse escapement from salmon cages is also a problem and efficacy with salmon larger than 2 kg requires further study. Better understanding of wrasse behaviour, including use of food and cover, in cages, and interactions with salmon, may facilitate improved wrasse husbandry, reduced escapement and improved cleaning efficiency.

13.1 Introduction

Cleaning behaviour by fish is a phenomemon frequently observed in tropical waters (Losey, 1972, 1974, 1987), but it has been less frequently observed in the north-east Atlantic. This absence of observation does not mean that cleaning is rare, but may reflect the more restricted visibility and lower numbers of observers swimming in these colder seas (Costello, 1991). The first observations of cleaning by north European fish were in aquaria (Potts, 1973; Samuelsen, 1981), and a few observations of cleaning in the sea (Potts, 1973; Hilldén 1983; author *unpubl. data*) indicated that cleaning was not an artefact of captivity. There are now records of corkwing [*Crenilabrus melops* (L.)], goldsinny [*Ctenolabrus rupestris* (L.)], rock cook [*Centrolabrus exoletus* (L.)] and female and juvenile cuckoo wrasse (*Labrus mixtus* L.), cleaning other fish species (Table 13.1).

The most commercially significant pathogens in fish culture in Europe are sea lice (Copepoda, Caligidae). Two species, *Lepeophtheirus salmonis* Krøyer and *Caligus elongatus* Nordmann, infest the skin of Atlantic salmon, *Salmo salar* L., farmed in sea cages in Norway, Scotland, Shetland, and Ireland (Pike, 1989). Not only are the hosts damaged such that their market value is reduced, but they are exposed to secondary infections for which the lice may act as vectors (Nylund *et al.*, 1991, 1993; Costello, 1993; Kvenseth, 1993). Both the lice and the secondary infections can cause major losses of salmon on the farms. The conventional treatment of lice involves bathing the salmon in dichlorvos and this is very stressful to the salmon (Salte *et al.*, 1987; Horsberg *et al.*, 1989; Høy *et al.*, 1991), costly, hazardous to farm staff (Health and Safety Executive, 1987), and only temporarily effective (Bron *et al.*, 1993a). This and other lice control methods were reviewed by Costello (1993).

The most innovative method of parasite control in fish farming has been the use of wrasse to clean lice off salmon. The basis of this cleaner-fish technology is that wrasse are captured, placed in salmon cages, and the wrasse swim amongst the circling salmon and pluck lice from their skin (Costello & Bjordal, 1990). The marine species used are goldsinny, rock cook, corkwing, and female cuckoo, and they are usually captured in modified shrimp pots or fyke nets. In this paper, the development of the use of cleaner-fish to control sea lice on Atlantic salmon in northern Europe is summarised. The lessons to be learned from this development and the potential of biological control to become more widespread in aquaculture is explored.

13.2   Sea lice control by wrasse on salmon farms

In Bergen, Norway, a specialist in fish behaviour, Åsmund Bjordal, decided to investigate the potential of cleaner-fish to control sea lice on salmon following inconclusive attempts by local fish farmers to see if wrasse would clean salmon. In 1987 he observed the cleaning of *L. salmonis* from Atlantic salmon in aquaria by locally captured wrasse (Labridae). These laboratory experiments were followed by trials in experimental cages and then commercial cages (Bjordal, 1988, 1990, 1991, 1992). In each instance, significant reductions of lice on the salmon occurred. From 1990, the use of wrasse to control sea lice on salmon farms (cleaner-fish

Table 13.1 Records of cleaning by wrasse in northern Europe including representative studies of cleaning in commercial salmon cages.

Cleaner species	Species cleaned	Location observed	Reference
CORKWING			
	Ballan[1]	Plymouth aquarium, UK	Potts 1973
	Plaice[2]	Plymouth aquarium	Potts 1973
	Black bream[3]	Plymouth aquarium	Potts 1973
	Red bream[4]	Plymouth aquarium	Potts 1973
	Mackerel[5]	Plymouth aquarium	Potts 1973
	Salmon[6]	Aquarium and sea cages, Bergen	Bjordal 1988
GOLDSINNY			
	Ballan	Plymouth aquarium	Potts 1973
	Ballan	Lough Hyne, Ireland	Darwall, Lysaght *pers. comm.*
	Ballan	Swedish coast	Hilldén 1983
	Ballan	South coast of England	Chapter 3, this volume
	Goldsinny	Plymouth aquarium	Potts 1973
	Halibut[7]	SFIA Aquarium, Ardtoe	Smith *pers. comm.*
	Plaice[2]	Plymouth aquarium	Potts 1973
	Salmon	Aquarium and sea cages, Bergen	Bjordal 1988
	Salmon	Sea cages, Shetland	Chapter 16, this volume
	Cuckoo	Sea near Flødivegen, Norway	author
	Lumpsucker[8]	Dunstaffnage Marine Laboratory	Sayer *pers. comm.*
	Red mullet[9]	South coast of England	Chapter 3, this volume
ROCK COOK			
	Ballan	Plymouth Sound, UK	Potts 1973
	Ballan	Lough Hyne, Ireland	Chapter 4, this volume
	Anglerfish[10]	Bergen aquarium	Samuelsen 1981
	Grey mullet[11]	Lough Hyne, Ireland	author
	Salmon	Aquarium and sea cages, Bergen	Bjordal 1988
	Salmon	Aquaria in Ireland	Tully *pers. comm.*
	Lumpsucker[8]	Dunstaffnage Marine Laboratory	Sayer *pers. comm.*
	Goldsinny	Aquarium, Bergen	Bjordal *pers. comm.*
FEMALE CUCKOO			
	Salmon	Aquarium and sea cages, Bergen	Bjordal 1988

Latin names (after Wheeler 1992): [1] *Labrus bergylta* Ascanius; [2] *Pleuronectes platessa* L.; [3] *Spondyliosoma cantharus* (L.); [4] *Pagellus bogareveo* (Brünnich); [5] *Scomber scombrus* L.; [6] *Salmo salar* L.; [7] *Hippoglossus hippoglossus* (L.); [8] *Cyclopterus lumpus* L.; [9] *Mullus surmuletus* L.; [10] *Lophius piscatorius* L.; [11] Mugilidae.

technology) became an established and increasingly important part of lice control measures in Norway (Kvenseth, 1993) (Table 13.2).

The initial laboratory and experimental studies in Norway were sponsored by the government at the Institute of Marine Research where Bjordal worked, and the commercial trials conducted in collaboration with a local fish farmer. The use of cleaner-fish continues to expand in Norway and Scotland (Table 13.2), with investment in research into wrasse fisheries and husbandry continuing in Norway.

Table 13.2   Summary of the historical development of cleaner-fish technology. Further details may be found in Bjordal (1988, 1990, 1991, 1992), Costello and Donnelly (1991), Darwall *et al.* (1992a), Treasurer (1991a, 1991b, 1993a, 1993b), Kvenseth (1993, Chapter 15, this vol.), and Chapter 16, this vol.

Year	Country	Development
1987	Norway	Discovery cleaner-fish could remove salmon lice
1988	Norway	First commercial trial is a success
1989	Shetland	First commercial trial is a success at Vaila Sound Salmon farm
1989	Scotland	First commercial trials (at Loch Sunart) not successful due to wrasse escapement
1990	Scotland	Successful commercial trials at Marine Harvest and Golden Sea Produce
1990	Ireland	First successful trial at Fanad
1991	Ireland	Successful trials at Fanad, Tully Mountain, Mhuir Gheal Teo, and Bradán Mhara conducted
1991	Norway	28 farms using 135 000 wrasse to clean 5.2 million salmon
1991	Scotland	Marine Harvest, Golden Sea Produce, McConnell Salmon, Highland Fish Farms, Strathaird Salmon, Kinloch Damph and Kerrera Fisheries all using cleaner-fish
1992	Ireland	Mhuir Gheal, Bradan Mhara, Killary Salmon using wrasse in more cages than 1991
1992	Scotland	Same farms and others expanding use of wrasse at their sites
1992	Norway	67 farms using 281 000 wrasse on 8.5 million salmon (including 937 500 second sea-year salmon)
1993	Ireland	Only Killary Salmon Ltd attempting to use wrasse
1993	Norway	137 farm sites using over 1 million wrasse on 17.2 million salmon
1994	Ireland	No farms using wrasse
1994	Scotland	30 farm sites using 150 000 wrasse
1994	Norway	130 farm sites using 1.5 million wrasse

The first use of wrasse in salmon cages outside Norway were initiated and funded entirely by commercial farms in Shetland (Chapter 16, this volume) and Ireland (Costello & Donnelly, 1991) in 1990. Subsequent trials and use of wrasse in Scotland and Shetland (Table 13.2) were funded by the industry and trials in Ireland by the industry alone, the government Sea Fisheries Development Board (Bord Iascaigh Mhara), and a research project funded by the Fisheries and Aquaculture Research programme of the Commission of the European Community (now the European Union). A survey of Scottish salmon farmers found wrasse to be the 'most popular' alternative to dichlorvos for sea lice control (Anon., 1991a).

The outcome of these trials was that cleaner-fish were shown to be commercially viable for the control of sea lice. They also proved to be significantly cheaper to use than dichlorvos (Anon., 1991b, 1993) and half the price of using hydrogen peroxide (Kvenseth, 1993). If the greater benefits in salmon health and growth rate with using wrasse (Costello *unpubl. data*) are also considered the economic savings are even greater. Additionally, in contrast to chemotherapeutants, wrasse are not hazardous to farm staff and do not stress the salmon. There may also be an added marketing advantage in that farmers have minimized or avoided the use of pesticides to control sea lice, and the industry is perceived to be reducing the release of pesticides into the marine environment.

In Ireland, cleaner-fish were not used by any farms in 1994 and the primary treatment for lice was the chemotherapeutant ivermectin. Ivermectin is effective in sea lice control and has advantages in being administered orally rather than as a bath (Smith *et al.*, 1993). However, because: (i) the manufacturer is unlikely to submit this compound for licensing for use in salmon farms (Brewer, 1991); (ii) there are no marine ecotoxicological data on the compound; (iii) it is persistent in the environment; (iv) it has a long withdrawal period in salmon; and (v), other salmon-producing countries have supported several other compounds but not ivermectin as a candidate for sea lice treatment on pharmacological grounds (Armstrong, 1994), it is likely that the Irish industry will need to reconsider other options, including the use of cleaner-fish, in the future.

Cleaner-fish technology is more similar to a management technique than convential chemotherapeutant methods of parasite control. The development of cleaner-fish and other management techniques to control sea lice, such as fallowing of cage sites (Grant & Treasurer, 1993), requires research funding from authorities and industry to demonstrate the effectiveness and limitations of the techniques, and industry involvement is essential. In contrast, development of chemotherapeutants, and to a lesser extent vaccines, can receive financial support from pharmaceutical companies as these form commercially marketable products.

The close involvement of salmon farmers with the development of cleaner-fish technology has been an important factor in the rapid integration of the technology in the salmon industry. It is notable that the scientists involved in the development of cleaner-fish technology have not been specialists in the treatment of fish parasites and diseases, but have backgrounds in fish behaviour (Bjordal), ecology (Costello), and fish biology (Treasurer). This exemplifies the value of a broad multidisciplinary approach to solving applied problems. From enquires I have made since 1988, of specialists in fish behaviour and disease treatment, as to the potential of cleaner-fish in controlling sea lice, I have received reactions ranging from scepticism to outright rejection of its feasibility. As many of these scientists were advisors to the salmon farming industry it is a credit to the open-minded attitude of many fish farmers that they supported and independently assessed the outcome of trials with cleaner-fish.

13.3   Ectoparasite infestations of other cultured fish

Ectoparasites are widespread on marine and freshwater fish, and it is not suprising that some parasites reach pathogenic levels when fish are cultured at high densities. It is likely that, as intensive finfish farming diversifies, more species will be affected by lice and other parasites. Cleaners are not particular about which species a parasite is, and cleaning of isopod and copepod crustaceans, trematode flatworms, and leeches has been observed (McCutcheon & McCutcheon, 1964;

Potts, 1973; Losey, 1974; Samuelsen, 1981; Gorlick *et al.*, 1987; Grover, 1994). Non-labrids may also have potential as cleaners, and species of goby, surfperch, pipefish and shrimp are known to show cleaning behaviour (Spall, 1970; Hobson, 1971; Losey, 1972; Potts, 1973; Losey, 1987; Johnson & Ruben, 1988). In the Bahamas, cleaning gobies (*Gobiosoma genie* Ginsburg), neon gobies [*G. oceanops* (Jordan)] and juvenile bluehead wrasse [*Thalasoma bifasciatum* (Bloch)], all cleaned trematode flatworms from tilapia cultured in seawater in experimental conditions (Cowell *et al.*, 1993; Grover, 1994) (Table 13.3).

Although cleaner-fish stations and host invitation postures are well documented for tropical waters, the experience in salmon farms is that neither of such behaviours is a prerequisite for cleaning. The potential of cleaner-fish is not always readily apparent, and in some countries cleaner-fish are not well known. Atlantic

Table 13.3 Other ectoparasites which have caused significant problems in aquaculture, and may be controlled by cleaner-fish (for further details see Costello 1993). Records of significant infestations on wild fish are also listed. All parasites are marine with the exception of the freshwater *Argulus foliaceus*.

Parasite	Host	Locality	Reference
On fish in culture			
Argulus foliaceus (L.)	rainbow brout[1]	Portugal	Menezes *et al.* 1990
Caligus elongatus Nordmann	red drum	USA	Landsberg *et al.* 1991
Caligus epidemicus Hewitt	tilapia[2]	Taiwan	Lin & Ho 1993
Caligus minimus (Otto)	sea-bass[3]	Greece	Papoutsoglou *et al.* 1995
Caligus orientalis Gussev	rainbow trout[1]	Japan	Urawa & Kato 1991
Caligus spinosus Yamaguti	yellow tail[4]	Japan	Izawa 1969
Pseudocaligus apodus Brian	mullet[5]	Israel	Paperna 1975
Ergasilus labracis Krøyer	Atlantic salmon[6]	Canada	O'Halloran *et al.* 1992
Ergasilus lizae Krøyer	mullet[5]	Middle East	Paperna 1975
Ergasilus lizae	mullet[5]	Mediteranean France	Ben-Hassine *et al.* 1983
Gnathia sp.	Atlantic salmon[6]	Ireland	Drinan & Rodger 1990
Neobenedenia melleni	tilapia[7]	Bahamas	Cowell *et al.* 1993, Grover 1994
Nerocila orbignyi Guérin-Meneville	sea-bass[3]	Corsica	Bragoni *et al.* 1983, 1984
On wild fish			
Argulus foliaceus	perch[8]	England	Pickering & Willoughby 1977, J.M. Elliott *pers. comm.*
Caligus elongatus	herring[9]	Scotland	MacKenzie & Morrison 1989
Caligus minimus	sea-bass[3]	Middle East	Paperna 1980
Caligus elongatus	cod[10], haddock[11]	N. Atlantic	Neilson *et al.* 1987
Lepeophtheirus salmonis Krøyer	sea-trout[12]	Ireland	Tully 1992, Tully & Whelan 1993, Tully *et al.* 1993a, 1993b

[1] *Oncorhynchus mykiss* (Walbaum); [2] *Oreochromis mossambicus* (Peters); [3] *Dicentrachus labrax* (L.); [4] *Seriola quinqueradiata* Temminck and Schlegel; [5] *Mugil cephalus* L.; [6] *Salmo salar L.;* [7] hybrid of *Oreochromis mossambicus* (Peters) and *O. urolepis hornorum* (Trewavas); [8] *Perca fluviatilus* L.; [9] *Clupea harengus* L.; [10] *Gadus morhua* L.; [11] *Melanogrammus aeglefinus* (L.); [12] *Salmo trutta* L.

salmon farms in Canada have been infested with both *Caligus elongatus* and *Lepeophtheirus salmonis* (Hogans & Trudeau, 1989; Stuart, 1990; Armstrong, 1994), but cleaning behaviour has not been observed in marine fish in Canadian waters. Several species of surfperches (Embiotocidae) known to show cleaning behaviour in Californian waters also occur in Pacific Canada (Hobson, 1971; Boschung *et al.*, 1983). There are two species of wrasse common in Atlantic Canada, the cunner, [*Tautolabrus adspersus* (Walbaum)] and tautog [*Tautoga onitis* (L.)] (Boschung *et al.*, 1983). In general appearance and ecology the cunner seems similar to the European ballan wrasse the adults of which do not show cleaning behaviour. Neither of these three labrids have the black spots or stripes typical of many other cleaner-fish and which may be a 'cleaner guild' identification mark (Eibl-Eibesfeldt, 1955). However, whether surfperches, cunner, tautog or other fishes (adult or juvenile) show cleaning behaviour in Candian waters has only begun to be tested (Mackinnon, 1995).

Ectoparasites are not equally available to cleaner-fish, and some parasites have adopted a lifestyle which places them out of reach of cleaner-fish. In the Mediterranean area, both wild and farmed sea-bass are infested by the lice species *Caligus minimus* (Table 13.3) and several species of cleaner-fish occur. However, *C. minimus* lives inside the opercula and mouth cavity and is thus unavailable to cleaner-fish. Most cleaner-fish will clean others of their species, and this probably explains the low numbers of ectoparasites on cleaning wrasse (this volume, Chapter 17). However, some parasites dwell within the skin of wrasse, such as the most common external parasite of corkwing, *Leposphilus labrei* Hesse (this volume, Chapter 17). Such cryptic ectoparasite lifestyles may have evolved as a response to cleaning.

Freshwater fish species from Europe, North America, Africa and Asia have been observed to act as cleaner-fish (Wickler, 1956; Spall, 1970; Abel, 1971; Wyman & Ward, 1972; Bartmann, 1973; Sulak, 1975; Able, 1976; Powell, 1984; Stauffer, 1991). Not only are freshwater parasites such as *Argulus* cleaned, but also bacterial, fungal and protozoan infections (Darkov & Panyushkin, 1988). Infestations of *Argulus* on rainbow trout in cages in a freshwater lake were so severe that the farm had to cease operation (Table 13.3). The collapse of a wild perch population in a lake was also probably caused by *Argulus*, although, because perch were collected in static nets, few *Argulus* were collected and this could not be confirmed. The general pattern in freshwater fish appears to be that cleaning is facultative, that smaller fish clean (a) larger fish of other species, and (b) similar-sized and larger individuals of their own species, and that once a skin infection has disappeared cleaning ceases. Typical host invitation postures are shown by carp (*Cyprinus carpio* L.) and Sumatran barbs [*Barbus tetrazona* (Bleeker)] (Darkov & Panyushkin, 1988; Soto *et al.*, 1994), while other hosts either stay still or move very slowly while being cleaned. Carp is one of the most widely cultured species in the world, and it may be particularly significant for its culture that intraspecific cleaning occurs in this species (Soto *et al.*, 1994).

13.4   Other applications of biological control in aquaculture

The growth of fouling organisms is a problem in both finfish and shellfish farming. In salmonid farming, fouling increases net weight, obscures net tears from the farmer, and decreases water flow. This results in increased net tear, and decreased oxygen replenishment and ammonia flushing from the cage. Farmers must therefore regularly (every few weeks) remove, clean and repair nets. Wrasse, added to salmon cages as cleaner-fish, browsed the cage fouling such that half the amount of net cleaning was required in Norwegian farms (Bjordal, 1992; Chapter 15, this volume). Some salmon farmers in Norway are also experimenting with sea-urchins to control cage fouling (Maroni, *pers. comm.*).

Fouling is a greater problem in shellfish farming compared to finfish farming, because the fouling organisms compete with the cultured species for space and food. The principle in biological control of shellfish fouling is to place locally-available predators of fouling organisms within the shellfish holding trays or baskets. These predators have to be selected by species and size to feed on the fouling without damaging the cultured shellfish. Biological control of fouling in shellfish culture has been practised since at least the early 1980s (Hidu *et al.*, 1981). Examples of the following have all been documented:

(a) killifish [*Fundulus heteroclitus* (L.)] controlling seasquirts [*Molgula manhattensis* (de Kay)] in clam [*Mercenaria mercenaria* (L.)] trays in New Jersey (Flimlin & Mathis, 1993);

(b) crabs [*Cancer irroratus* Say, *Carcinus maenas* (L.), *Pagurus acadianus* (Benedict)] controlling mussels, seasquirts, and sponges fouling oyster (*Ostrea edulis* L.), trays in Nova Scotia (Enright *et al.*, 1993);

(c) periwinkles [*Littorina littorea* (L.)], controlling algae (*Ectocarpus*, *Enteromorpha*, and *Ulva* spp.) fouling in oyster culture (Enright *et al.*, 1983; Enright, 1993); and

(d) dogwhelks [*Nucella lapillus* (L.)] controlling mussels (*Mytilus edulis* L.) on scallop [*Pecten maximus* (L.)] and oyster trays in Ireland (Minchin & Duggan, 1989).

As with cleaner-fish, biological control of fouling organisms is a cheap, natural and environmentally benign technique, which is neither hazardous to the farmer or the environment in contrast to the use of chemotherapeutants. As most farmers are already familiar with marine life and handling fish, they can readily manage and adapt biological control techniques.

Following some initial inertia, biological control is now an integral part of agriculture and has significant economic benefits to farmers (Chant, 1966; Lee, 1981), without considering the additional environmental benefits in reduced use of biocides. The more rapid adoption of biological control in aquaculture than agriculture reflects the rapid rate of growth and innovation in modern fish farming, but may also have indirectly benefited from the acceptance of the principle of biological control in agriculture.

13.5   Limitations of cleaner-fish technology

Every technology has its limitations, which if recognized may be overcome or minimized by further research. The main limitations in the further use of cleaner-fish to control salmon lice are: wrasse availability; concerns over transfer of pathogens from wrasse to farmed salmon; efficacy with large (>2 kg) second sea-year salmon; escapement from cages; and scepticism and misinformation about the effectiveness of cleaner-fish.

The development of cleaner-fish technology has created a new fishery for wrasse (Treasurer & Henderson, 1992). However, the sustainability of localized fisheries for wrasse large enough to be contained within salmon cages (>10 cm in body length) is limited (Darwall *et al.*, 1992b). Although abundant in southern Norway, wrasse abundance is more patchy in northern Norway, Scotland and Ireland, and insufficient wrasse are available in Shetland to supply the salmon farms (Costello, *unpubl. data*). With the exception of southern Norway, some farms will not have sufficient wrasse in their immediate vicinity and would have to import wrasse.

The transfer of wild wrasse from one area to another raises concerns about disease transfer between areas. Indeed this is a concern whether the disease affects the farmed salmon or not. While it is recommended that wrasse from all populations used as cleaner-fish are screened for pathogens, this is particularly important for populations from which wrasse are being transferred between areas. Even with this screening, it is difficult to detect parasites or diseases present at low frequencies in the population, and screening for viruses and protozoans is technically difficult (Chapter 17, this volume). A possible solution to this problem of supply and disease free status would be to culture wrasse. While this appears feasible (Anon., 1993; this volume, Chapter 12), commercial scale production techniques have not been demonstrated. For their weight, wrasse (at £0.5–1.5 each) are probably the most valuable fish in Europe. Culturing wrasse would be profitable at present if conducted within an existing commercial marine fish culture facility (Barbour & Costello, *unpubl. data*), but there are few such facilities in northern Europe.

Another aspect to wrasse availability has been that the fishery is limited to the late spring to early autumn months, generally June to October inclusive, due to the temperature-dependent activity of wrasse (Darwall *et al.*, 1992b; Sayer *et al.*, 1993; Costello *et al.*, 1995; this volume, Chapter 21). Wrasse are required by salmon farmers when smolts are put to sea in April to May. However, this mismatch between demand and fishery may be overcome by the overwintering of wrasse in rigid cages sunk to the seabed, as demonstrated in Norway (Skog *et al.* 1994; Chapter 22, this volume).

Some farmers have successfully controlled lice with cleaner-fish on salmon from smolt to harvest two years later (Thorburn, 1991; Chapter 16, this volume). However, large salmon can attack and eat wrasse (Anon., 1993; Kvenseth, 1993). While this problem may be avoided if the salmon have been stocked with wrasse since first put into the sea, there is insufficient understanding of behavioural interactions between wrasse and salmon to determine how to minimize this problem. Similarly, further understanding of wrasse behaviour and needs in cages may help to reduce their escape. Escape from cages is significant even within mesh sizes through which wrasse cannot escape. Evidently the wrasse take advantage of even the smallest tear in a net to escape. The use of hides or provision of food for wrasse in the cages may reduce escapement but the benefits of such husbandry techniques remain to be quantified.

Rumours that cleaner-fish were not effective, ate salmon food in preference to sea lice, or attacked salmon eyes causing blinding and death, have found no foundation in documented trials. The development of such rumours was perhaps assisted by the speed of adoption of the use of cleaner-fish, such that the available data tended to be anecdotal or qualitative. It is also difficult to conduct experiments with adequate controls on commercial farms. However, trials in different countries and farms within countries have repeatedly demonstrated effective lice control by cleaner-fish, and considerable data has now been published (Bjordal, 1990, 1991, 1992; Costello & Donnelly, 1991; Treasurer, 1991a, 1991b, 1993a, 1993b; this volume, Chapter 14; Kvenseth, 1993, Chapter 15, this volume).

Already, cleaner-fish technology is an established method of sea lice control in Atlantic salmon farming in Europe. The use of cleaner-fish compliments other lice control strategies, notably avoiding overlapping generations of salmon in a bay, fallowing sites and bays, and treatment with chemotherapeutants if necessary (Bron *et al.*, 1993b; Grant & Treasurer, 1993). The future development of cleaner-fish technology would benefit most by: first, the supply of cultured and disease-free certified wrasse; second, improved wrasse husbandry through better understanding of wrasse behaviour in nature and salmon cages; and lastly improved understanding of wrasse parasites and diseases.

Acknowledgements

This study was part-funded by the European Union under Fisheries and Aquaculture Research programme contract AQ.2.502 on cleaner-fish technology to the author. I thank my colleagues in this project, and others, notably A. Barbour, Å. Bjordal, P. Kvenseth, M. Sayer, J. Treasurer, C. Young, for their helpful and enthusiastic discussion.

References

Abel, E.F. (1971) Zur Ethologie von Putzsymbiosen einheimscher Süsswasserfische im natürlichen Biotop. *Oecologia (Berlin)* **6**: 133–51.

Able, K.W. (1976) Cleaning behaviour in the cyprinodontid fishes: *Fundulus majalis*, *Cyprinodon variegatus*, and *Lucania parva*. *Chesapeake Science* **17**: 35–9.

Anon. (1991a) Scottish farmers take long view on louse control. *Fish Farmer* **14**: 45.

Anon. (1991b) Wrasse halve expenses. *Fish Farming International* **18(11)**: 46.

Anon. (1993) Hydro companies test farming of cleaner fish. *Fish Farming International* **20(1)**: 12–13.

Armstrong, R. (1994) Sea lice treatment registration for Canadian fish farms. *Bulletin of the Aquaculture Association of Canada* **94–1**: 29–33.

Bartmann, W. (1973) Eine Putzsymbiose zwischen Stichling (*Gasterosteus aculeatus* L.) und Hecht (*Esox lucius* L.) im Aquarium. *Zeitschrift für Tierpsychologie* **33**: 153–62.

Ben-Hassine, O.K., Braun, M. and Raibaut, A. (1983) Comparative study of the infestation of *Mugil cephalus* L., 1758 by the copepod *Ergasilus lizae* Kroyeri, 1863 in two lagoons of the French Mediterranean coast. *Rapport P.V. Ren. CIESM* **28(6)**: 379–84.

Bjordal, Å. (1988) Cleaning symbiosis between wrasses (Labridae) and lice infested salmon (*Salmon salar*) in mariculture. *International Council for the Exploration of the Sea*, Mariculture Committee, 1988/F **17**: 8 pp.

Bjordal, Å. (1990) Sea lice infestation of farmed salmon: possible use of cleaner fish as an alternative method for delousing. *Canadian Technical Reports in Fisheries and Aquatic Sciences* **1761**: 85–9.

Bjordal, Å. (1991) Wrasse as cleaner fish for farmed salmon. *Progress in Underwater Science* **16**: 17–29.

Bjordal, Å. (1992) Cleaning symbiosis as an alternative to chemical control of sea lice infestation of Atlantic salmon. In *The Importance of Feeding Behaviour for the Efficient Culture of Salmonid Fishes* (Ed. by Thorpe, J.E. and Huntingford, F.A.). World Aquaculture Workshops, No. 4, World Aquaculture Society, pp. 53–60.

Boschung, H.T., Williams, J.D., Gotshall, D.W., Caldwell, D.K. and Caldwell, M.C. (1983) *The Audubon Society field guide to North American Fishes, Whales, and Dolphins*. A.A. Knopf, New York.

Bragoni, G., Romestand, B. and Trilles, J.P. (1983) Parasitism by cymothoids among bass (*Dicentrarchus labrax* L., 1758) in rearing. II. Parasitic ecophysiology in Diana Pond (Corsica). *Annales Parasitologie Hum. Comp.* **58**: 593–609.

Bragoni, G., Romestand, B., and Trilles, J.-P. (1984) Parasitoses a cymothoadien chez le loup, *Dicentrarchus labrax* (Linnaeus, 1758) en élevage. I. Écologie parasitaire dans le cas de l'étang de Diana (Haute-Corse) (Isopoda, Cymothoidae). *Crustaceana* **47**: 44–51.

Brewer, M.F. (1991) Letter to the editor. *Fish Farmer* **14(5)**: 7.

Bron, J.E., Sommerville, C., Wootten, R. and Rae, G.H. (1993a) Fallowing of marine Atlantic salmon, *Salmo salar* L., farms as a method for the control of sea lice, *Lepeophtheirus salmonis* (Krøyer, 1837). *Journal of Fish Diseases* **16**: 487–93.

Bron, J.E., Sommerville, C., Wootten, R. and Rae, G.H. (1993b) Aspects of the epidemiology of the sea lice *Lepeophtheirus salmonis* (Krøyer, 1837) and *Caligus elongatus* Nordmann, 1832 affecting farmed salmon, *Salmo salar* Linnaeus, 1758 on the west coast of Scotland. In *Pathogens of Wild and Farmed Fish: Sea Lice* (Ed. by Boxshall, G.A. and Defaye, D.). Ellis Horwood Ltd., Chicester, pp. 263–74.

Chant, D.A. (1966) Integrated control systems. In *Scientific aspects of pest control*. National Academy of Sciences, National Research Council, Publ. No. 1402, pp. 193–218.

Costello, M.J. (1991) Review of the biology of wrasse (Labridae: Pisces) in Northern Europe. *Progress in Underwater Science* **16**: 29–51.

Costello, M.J. (1993) Review of methods to control sea lice (Caligidae, Crustacea) infestations on salmon farms. In *Pathogens of Wild and Farmed Fish: Sea Lice* (Ed. by Boxshall, G.A. and Defaye, D.). Ellis Horwood Ltd., Chicester, pp. 219–52.

Costello, M.J. and Bjordal, Å. (1990) How good is this natural control of sea lice? *Fish Farmer* **13(3)**: 44–6.

Costello, M.J., and Donnelly, R. (1991) Development of wrasse technology. In *Bradán '90* (Ed. by Joyce, J.R.). Irish Salmon Growers' Association, Dublin, pp. 18–20.

Costello, M.J., Lysaght, S. and Darwall, W.R. (1995) Temporal variation in the abundance of north European wrasse as estimated by diver observations. In *Biology and Ecology of Shallow Coastal Waters* (Eds A. Eleftherion, A.D. Ansell & C.Y. Smith) Proceedings of the 28th European Marine Biology Symposium, Crete 1993, Olsen and Olsen, Fredenburg pp. 343–50.

Cowell, L.E., Watanabe, W.O., Head, W.D., Grover, J.J. and Shenker, J.M. (1993) Use of tropical cleaner fish to control the ectoparasite *Neobenedenia melleni* (Monogenea: Capsalidae) on seawater-cultured Florida red tilapia. *Aquaculture* 113: 189–200.

Darkov, A.A. and Panyushkin, S.N. (1989) Cleaning symbioses in six freshwater fish species. *Journal of Ichthyology* 28: 161–7.

Darwall, W., Costello, M.J. and Lysaght, S. (1992a) Wrasse–how well do they work? *Aquaculture Ireland* 50: 26–9.

Darwall, W.R.T., Costello, M.J., Donnelly, R., and Lysaght, S. (1992b) Implications of life-history characteristics for a new wrasse fishery. *Journal of Fish Biology* 41 (Supplement B): 111–23.

Drinan, E.M. and Rodger, H.D. 1990. An occurrence of *Gnathia* sp., ectoparasitic isopods, on caged Atlantic salmon. *Bulletin of the European Association of Fish Pathologists* 10: 141–2.

Eibl-Eibesfeldt, I. (1955) Über Symbiosen, Parasitismus und andere besondere zwischenartliche Beziehungen tropischen Meererfische. *Zeitschrift für Tierpsychologie* 12: 204–19.

Enright, C.T. (1993) Control of fouling in bivalve aquaculture. *World Aquaculture* 24: 44–6.

Enright, C.T., Elner, R.W., Griswold, A. and Borgese, E.M. (1993) Evaluation of crabs as control agents for biofouling in suspended culture of European oysters. *World Aquaculture* 24: 49–51.

Enright, C., Krailo, D., Staples, L., Smith, M., Vaughan, C. Ward, D., Gaul, P. and Borgese, E. (1983) Biological control of fouling algae in oyster culture. *Journal of Shellfish Research* 3: 41–4.

Flimlin, G.E. and Mathis, G.W. Jr (1993) Biological fouling control in a field based nursery for the hard clam, *Mercenaria mercenaria*. *World Aquaculture* 24: 47–8.

Grant, A.N. and Treasurer, J.W. (1993) The effects of fallowing on caligid infestations on farmed Atlantic salmon (*Salmo salar* L.) in Scotland. In *Pathogens of Wild and Farmed Fish: Sea Lice* (Ed. by Boxshall, G.A. and Defaye, D.). Ellis Horwood Ltd, Chichester, pp. 255–60.

Gorlick, D.L., Atkins, P.D. and Losey, G.S. (1987) Effect of cleaning by *Labroides dimidiatus* (Labridae) on an ectoparasite population infecting *Pomatocentrus vaiuli* (Pomatocentridae) at Enewetak Atoll. *Copeia* 1987: 41–5.

Grover, J. (1994) A new role for cleaner fish: biological control of ectoparasites in aquaculture. *Underwater Naturalist* 22: 9–13.

Health and Safety Executive (1987) *Biological Monitoring of Workers Exposed to Organophosphorus Pesticides*. Guidance note MS17, Dept. Environment, London.

Hidu, H., Conary, C. and Chapman, S. (1981) Suspended culture of oysters: biological fouling control. *Aquaculture* 22: 189–92.

Hilldén, N.-O. (1983) Cleaning behaviour of the goldsinny (Pisces: Labridae) in Swedish waters. *Behavioural Processes* 8: 87–90.

Hobson, E.S. (1971) Cleaning symbiosis among Californian inshore fishes. *Fishery Bulletin* 69: 491–523.

Hogans, W.E. and Trudeau, D.J. (1989) Preliminary studies on the biology of sea lice, *Caligus elongatus, Caligus curtus* and *Lepeophtheirus salmonis* (Copepoda: Caligoida) parasitic on cage-cultured salmonids in the lower Bay of Fundy. *Canadian Technical Reports in Fisheries and Aquatic Sciences* 1715.

Horsberg, T.E., Høy, T., and Nafstad, I. (1989) Organophosphate poisoning of Atlantic salmon in connection with treatment against salmon lice. *Acta Veterinaria Scandinavica* 30: 385–90.

Høy, T., Horsberg, T.E., and Wichstrøm, R. (1991) Inhibition of acetylcholinesterase in rainbow trout following dichlorvos treatment at different water oxygen levels. *Aquaculture* 95: 33–40.

Izawa, K. (1969) Life history of *Caligus spinosus* Yamaguti, 1939 obtained from cultured yellow tail, *Seriola quinqueradiata* T. and S. (Crustacea, Caligoida). *Report of Faculty of Fisheries, Prefectural University of Mie* 6: 127–57.

Johnson, W.S. and Ruben, P. (1988) Cleaning behaviour of *Bodianus rufus, Thalassoma bifasciatum, Gobiosoma evelynae*, and *Periclimenes pedersoni* along a depth gradient at Salt River Submarine Canyon, St Croix. *Environmental Biology of Fishes* 3: 225–32.

Kvenseth, P.G. (1993) Use of wrasse to control salmon lice. In *Fish farming technology* (Ed. by Reinertsen, H., Dahle, L.A., Jørgensen, L. and Tvinnerein, K.). Balkema, Rotterdam, pp. 227–32.

Landsberg, J.H., Vermeer, G.K., Richards, S.A., and Perry, N. (1991) Control of the parasitic copepod *Caligus elongatus* on pond-reared red drum. *Journal of Aquatic Animal Health* 3: 206–9.

Lee, B. (1981) Pests control pests: but at what price? *New Scientist* **89**, 150–2.

Lin, C.-L., and Ho, J. (1993) Life history of *Caligus epidemicus* Hewitt parasitic on tilapia (*Oreochromis mossambicus*) cultured in brackish water. In *Pathogens of Wild and Farmed Fish: Sea Lice* (Ed. by Boxshall, G.A. and Defaye, D.). Ellis Horwood Ltd, Chicester, pp. 5–15.

Losey, G.S. Jr. (1972) The ecological importance of cleaning symbiosis. *Copeia* **1972**: 820–33.

Losey, G.S. Jr. (1974) Cleaning symbiosis in Puerto Rico with comparison to the tropical Pacific. *Copeia* **1974**: 960–70.

Losey, G.S. Jr. (1987) Cleaning symbiosis. *Symbiosis* **4**: 229–58.

MacKenzie, K. and Morrison, J.A. (1989) An unusually heavy infestation of herring (*Clupea harengus* L.) with the parasitic copepod *Caligus elongatus* Nordmann, 1832. *Bulletin of the European Association of Fish Pathologists* **9**: 12–13.

MacKinnon, B.H (1995) The poor potential of cunner, *Tautogolabrus adspersus*, to act as cleaner-fish in removing sea lice (*Caligus elongatus*) from farmed salmon in eastern Canada. *Canadian Journal of Fisheries and Aquatic Sciences* **52** (Supplement 1): 175–7.

McCutcheon, F.H. and McCutcheon, A.E. (1964) Symbiotic behaviour among fishes from temperate ocean waters. *Science* 948–9.

Menezes, J., Ramos, M.A., Pereira, T.G. and da Silva, A.M. (1990) Rainbow trout culture failure in a small lake as a result of massive parasitosis related to careless fish introductions. *Aquaculture* **89**: 123–6.

Minchin, D. and Duggan, C.B. (1989) Biological control of the mussel in shellfish culture. *Aquaculture* **81**: 97–100.

Neilson, J.D., Perry, R.I., Scott, J.S. and Valerio, P. (1987) Interactions of caligid ectoparasites and juvenile gadids on Georges Bank. *Marine Ecology Progress Series* **39**: 221–32.

Nylund, A., Bjørknes, B. and Wallace, C. (1991) *Lepeophtheirus salmonis*–a possible vector in the spread of diseases on salmonids. *Bulletin of the European Association of Fish Pathologists* **11**: 213–16.

Nylund, A., Wallace, C. and Hovland, T. 1993. The possible role of *Lepeophtheirus salmonis* in the transmission of infectious salmon anaemia. In *Pathogens of Wild and Farmed Fish: Sea Lice* (Ed. by Boxshall, G.A. and Defaye, D.). Ellis Horwood Ltd, Chicester, pp. 367–73.

O'Halloran, J., Carpenter, J., Ogden, D., Hogans, W.E. and Jansen, M. (1992) *Ergasilus labracis* on Atlantic salmon. *Canadian Veterinary Journal* **33**: 75.

Paperna, I. (1975) Parasites and diseases of the grey mullet (Mugilidae) with special reference to the seas of the near east. *Aquaculture* **5**: 65–80.

Paperna, I. (1980) Study of *Caligus minimus* (Otto, 1821) (Caligidae, Copepoda) infections of the sea bass (*Dicentrarchus labrax* L.) in Bardawil lagoon. *Annales de Parasitologie* **55**: 687–706.

Papoutsoglou, S., Costello, M.J., Stamou, E. and Tziha, G. (1995) Environmental conditions at sea-cages, and ectoparasites on bass, *Dicentrarchus labrax*, and bream *Sparus aurata*, at two farms in Greece. *Aquaculture and Fisheries Management* (in press).

Pickering, A.D., and Willoughby, L.G. (1977) Epidermal lesions and fungal infection on the perch, *Perca fluviatilis* L., in Windermere. *Journal of Fish Biology* **11**: 349–354.

Pike, A.W. (1989) Sea lice–major pathogens of farmed Atlantic salmon. *Parasitology Today* **5**: 291–7.

Potts, G.W. (1973) Cleaning symbiosis among British fish with special reference to *Crenilabrus melops* (Labridae). *Journal of the Marine Biological Association of the United Kingdom* **53**: 1–10.

Powell, J.A. (1984) Observations of cleaning behaviour in the bluegill (*Lepomis macrochirus*), a centrarchid. *Copeia* **1984**: 996–8.

Salte, R., Syvertsen, C., Kjønnøy, M. and Fonnum, F. (1987) Fatal acetycholinesterase inhibition in salmonids subjected to a routine organophosphate treatment. *Aquaculture* **61**: 173–9.

Samuelsen, T.J. (1981) Der Seeteufel (*Lophius piscatorius* L.) in Gefangenschaft. *Zeitschrift Kolner Zoo* **24**: 17–19.

Sayer, M.D.J., Gibson, R.N. and Atkinson, R.J.A. (1993) Distribution and density of populations of goldsinny wrasse (*Ctenolabrus rupestris*) on the west coast of Scotland. *Journal of Fish Biology* **43** (Supplement A): 157–67.

Skog, K., Mikkelsen, K.O. and Bjordal, Å. (1994) Overvintring av bergnebb (*Ctenolabrus rupestris*) i lukkede teiner. *Fisken og havet* **5**, 1–5.

Smith, P.R., Maloney, M., McElligott, A., Clarke, S., Palmer, R., O'Kelly, J. and O'Brien, F. (1993)

The efficiency of oral ivermectin in the control of sea lice infestations of farmed Atlantic salmon. In *Pathogens of Wild and Farmed Fish: Sea Lice* (Ed. by Boxshall, G.A. and Defaye, D.). Ellis Horwood Ltd, Chichester, pp. 296–307.

Soto, C.G., Zhang, J.S. and Shi, Y.H. (1994) Intraspecific cleaning behaviour in *Cyprinus carpio* in aquaria. *Journal of Fish Biology* **44**: 172–4.

Spall, R.D. (1970) Possible cases of cleaning symbiosis among freshwater fishes. *Transactions of the American Fisheries Society* **99**: 599–600.

Stauffer, J.R. Jr. (1991) Description of a facultative cleanerfish (Telostei: Cichlidae) from Lake Malawi, Africa. *Copeia* **1991**: 141–7.

Stuart, R. (1990) Sea lice, a maritime perspective. *Aquaculture Association of Canada Bulletin* **1990**: 18–24.

Sulak, K.J. (1975) Cleaning behavior in the centrarchid fishes, *Lepomis macrochirus* and *Micropterus salmoides*. *Animal Behaviour*. **23** 331–4.

Thorburn, I. (1991) Green–the way to go. *Fish Farmer* **14(2)**: 46.

Treasurer, J. (1991a) Wrasse need due care and attention. *Fish Farmer* **14(4)**: 24–6.

Treasurer, J. (1991b) Limitations in the use of wrasse. *Fish Farmer* **14(5)**: 12–13.

Treasurer, J.W. (1993a) More facts on the role of wrasse in louse control. *Fish Farmer* **16(1)**: 37–8.

Treasurer, J.W. (1993b) Management of sea lice (Caligidae) with wrasse (Labridae) on Atlantic salmon (*Salmo salar* L.) farms. In *Pathogens of Wild and Farmed Fish: Sea Lice* (Ed. by Boxshall, G.A. and Defaye, D.). Ellis Horwood Ltd, Chicester, pp. 335–45.

Treasurer, J. and Henderson G. (1992) Wrasse: a new fishery. *World Fishing* **41**: 2.

Tully, O. (1992) Predicting infestation parameters and impacts of caligid copepods in wild and cultured fish populations. *Invertebrate Reproduction and Development* **22**: 91–102.

Tully, O. and Whelan, K.F. (1993) Production of nauplii of *Lepeophtheirus salmonis* (Krøyer) (Copepoda: Caligidae) from farmed and wild salmon and its relation to the infestation of wild sea trout (*Salmo trutta* L.) off the west coast of Ireland in 1991. *Fisheries Research* **17**: 187–200.

Tully, O., Poole, W.R., and Whelan, K.F. (1993a) Infestation parameters for *Lepeophtheirus salmonis* (Krøyer) (Copepoda: Caligidae) parasitic on sea trout, *Salmo trutta* L., off the west coast of Ireland during 1990 and 1991. *Aquaculture and Fisheries Management* **24**: 545–55.

Tully, O., Poole, W.R., Whelan, K.F., and Merigoux, S. (1993b) Parameters and possible causes of epizootics of *Lepeophtheirus salmonis* (Kröyer) infesting sea trout (*Salmo trutta* L.) on the west coast of Ireland. In *Pathogens of Wild and Farmed Fish: Sea Lice* (Ed. by Boxshall, G.A. and Defaye, D.). Ellis Horwood Ltd, Chichester, pp. 202–13.

Urawa, S., and Kato, T. (1991) Heavy infection of *Caligus orientalis* (Copepoda: Caligidae) on caged rainbow trout *Oncorhynchus mykiss* in brackish water. *Gyobyo Kenkyu* **26**: 161–2.

Wheeler, A. (1992) A list of the common and scientific names of fishes of the British Isles. *Journal of Fish Biology* **41** (Supplement A): 37 pp.

Wickler, W. (1956) Eine Putzsymbiose zwischen *Corydoras* und *Trichogaster*. *Zeitschrift für Tierpsychologie* **13**: 46–9.

Wyman, R.L. and Ward, J.A. (1972) A cleaning symbiosis between the cichlid fishes *Etroplus maculatus* and *Etroplus suratensis*. I. Description and possible evolution. *Copeia* **1972**: 834–8.

Chapter 14

Wrasse (Labridae) as cleaner-fish of sea lice on farmed Atlantic salmon in west Scotland

J.W. TREASURER *Marine Harvest McConnell, Lochailort, Inverness-shire PH38 4LZ, UK*

Wrasse were first used as cleaner-fish of sea lice on farmed salmon in Scotland on a trial basis from 1989 and commercially from 1990. The results of trials using wrasse and the extent of stocking with wrasse on salmon farms is reviewed. Wrasse controlled numbers of *Lepeophtheirus salmonis* in salmon in the first sea year until December but were less effective cleaners of *Caligus elongatus*. Wrasse showed less cleaning activity when stocked later in the production cycle and through the winter. From a questionnaire it was calculated that 34% of salmon farms on the Scottish mainland and Western Isles have used wrasse in the past five years. In 1994, 150 000 wrasse were stocked, representing 39% of smolts put to sea in that year in Scotland. Several problems with wrasse use still exist and guidelines for optimum application, particularly on the importance of suitable hides, are given.

14.1 Experiences with wrasse in Scotland

Wrasse were first used in Scotland in 1989 by the Sea Fish Industry Authority with trials in Shetland (Chapter 16, this volume) and in a small trial in Loch Sunart. The latter was hampered by cormorants *Phalacrocorax carbo* (L.) picking wrasse from the mortality collection sock and wrasse were lost as the salmon net mesh size was too large. Fish farmers stocked wrasse in Scotland, mainly goldsinny, *Ctenolabrus rupestris* (L.) and rock cook, *Centrolabrus exoletus* (L.), from 1990 on a commercial scale although mainly on an experimental basis. Initially many farmers tried to catch their own wrasse but they did not have the time to do this and numbers obtained were only sufficient to stock a few cages. Golden Sea Produce was perhaps the first company to ask local fishermen to supply wrasse and, in 1990, 35 000 wrasse were stocked with the 1990 salmon year class at a ratio of 1:50 salmon (Rae, *unpubl. data*). Treatments with dichlorvos were reduced and, on one farm, there was no requirement for chemical treatments to control lice.

In reviewing the use of wrasse in Scotland it has been difficult to obtain quantitative information on the relative numbers of lice in cages with and without wrasse as several farmers stocked wrasse in all cages, or records were not kept.

However, it became apparent that wrasse worked well in reducing sea lice numbers with salmon in their first sea-year. For example, at Loch Torridon, goldsinny and rock cook wrasse were stocked in a cage at a ratio of 1:100 salmon in August (Treasurer, 1991). Three treatments with dichlorvos were given. In the first six weeks of the trial there were no noticeable differences in lice numbers between this cage and two control cages. It was concluded that wrasse required an initial period to become trained in eating lice. From October, the cage with wrasse was not treated with dichlorvos and numbers of *Lepeophtheirus salmonis* Krøyer remained in the range of three to six mobiles through to January. During the same period, the two control cages were treated with dichlorvos on three occasions. Lice numbers increased in February in the trial cage because wrasse either escaped or, as reported in the Shetland trials (Chapter 16, this volume), showed reduced cleaning activity at reduced temperatures.

A more detailed trial was conducted with post-smolts at Loch Eil in 1991 in an experimental group of 5 m² cages located on a commercial farm with 75 000 fish of the 1990 year class (Treasurer, 1993). Wrasse were stocked in 18 cages while four cages were chosen at random on the group and had no wrasse. The mean number of lice (all *L. salmonis*) in the cages with wrasse fluctuated from three to twelve from 28 August 1991 to February 1992 (Fig. 14.1). Lice numbers on the four control cages rose to a maximum of 46 per fish by 14 October 1991. Wrasse were transferred to these cages to avoid treating them with dichlorvos and there was an immediate reduction in numbers of lice. When wrasse were removed and returned to their original cages after eight days, the numbers of lice rose to 45 per fish by November. Wrasse selected the larger lice developmental stages, mainly pre-adult

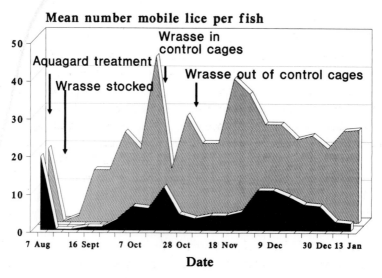

Fig. 14.1 A comparison of numbers of mobile *L. salmonis* on salmon from cages with (black, *n* = 18) and without wrasse (hatched, *n* = 4) at Loch Eil, 1991.

No. mobile lice, geometric mean

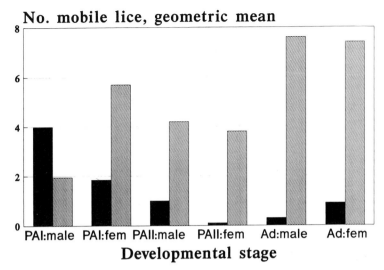

Developmental stage

Fig. 14.2 The composition of developmental stages of *L. salmonis* in cages with (solid) and without (hatched) wrasse. The mean number of all lice developmental stages was 7.4 per fish in cage 13 (solid) and 30.3 in cage 15 (hatched). The total length (mm) of each developmental stage ($n = 10$) measured in this study was PA I males 3.2–3.6, PA I females 3.1–4.1, PA II males 4.3–4.9, PA II females 5.0–5.7, adult males 5.2–5.9, adult females 8.5–11.8.

stage 2 (PAII) females and adults of both sexes (Fig. 14.2) (Treasurer, 1993). Adults comprised 6% of all lice present in cages with wrasse, compared with 48% on fish in control cages. The number of lice consumed by an individual goldsinny was in the range 26 to 46 lice day^{-1} and wet weight 349 to 907 mg, representing 1.2 to 2.7% of body weight for a fish of 30 g (Treasurer, 1994). This was similar to the cleaning rates of goldsinny of 28 to 45 lice in two trials in aquaria and up to 20 lice in a sea cage (Bjordal, 1991).

These results were not reproduced with larger one sea winter fish of *c.* 3 kg. In Loch Sunart wrasse were stocked in three cages on two farms on 5 August 1990 at ratios of 1:25, 1:50 and 1:100 respectively with *c.* 10 000 salmon in each 16 m^2 cage. The fourth cage in the block of four on each farm had no wrasse and acted as a control. Lice numbers rose to 100–200 per fish and treatment with dichlorvos was necessary within three weeks of stocking wrasse and two further treatments were required in September. There was no evidence of successful cleaning activity and it has generally been found that wrasse have to be introduced with salmon from the beginning of the production cycle (This volume, Chapter 16). Recruitment of lice was high throughout the trial and the salmon were too large. On harvesting only two wrasse were recovered. As the salmon net mesh size was 12 mm, wrasse of >100 mm should have been retained (Treasurer, 1991). It is possible that wrasse were eaten by salmon as they grew or escaped through small holes in the cage netting. This is the only documented trial in Scotland where different stocking densities of wrasse were examined. Later, from general experience, it has been found that a stocking ratio of 1:50 is more effective than 1:100.

The use of wrasse with salmon broodstock has been more encouraging. At Loch Ailort wrasse were introduced to salmon of 6.5 kg mean weight in 10 × 7 m cages. There were only 360 salmon in the cages and it was possible to use high stocking densities of wrasse of 1:10. The fish were treated with dichlorvos in August prior to the trial commencing. The numbers of mobile *L. salmonis* in the cage with wrasse declined from 10 in August to 3 in November, compared with an increase to 18 in the adjacent cage without wrasse (Fig. 14.3). Several factors may have contributed to the success of this trial: *viz* the high stocking ratio of wrasse; there was no mortality collecting sock fitted in these nets; the salmon had a diminished appetite as they were mature fish; the salmon were slow swimming, therefore making it easier for wrasse to clean lice. The mortality sock is a bag in the centre of the net bottom designed to hold fish mortalities and, where this is present, wrasse frequently use it as a hide because of its deeper/darker location and shelter provided.

A trial comparing the effectiveness of different wrasse species, namely goldsinny, rock cook, corkwing [*Crenilabrus melops* (L.)], and cuckoo wrasse (*Labrus mixtus* L.), was conducted on a farm in Loch Linnhe with first sea-year salmon. Unfortunately, the wrasse were stocked on 7 December 1990, too late in the year, and no useful results were obtained. Farmers in Scotland have stocked all species but not ballan wrasse (*Labrus bergylta* Ascanius) and most report that goldsinny and rock cook are mainly used. No comparative data on the cleaning ability of these species have been reported from Scotland.

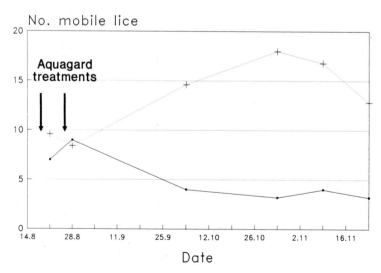

Fig. 14.3 Lice numbers in cages with salmon broodstock of 6.5 kg, with (solid line) and without (dotted line) wrasse stocked at a ratio of 1:10 salmon. Arrows indicate dates of treatment with dichlorvos.

After fallowing farms, *Caligus elongatus* Nordmann can be the principal sea louse species in the first year of the production cycle. Evidence on the effectiveness of wrasse against this species in Scotland is variable. In Loch Sunart wrasse were stocked in all cages in 1991 at a ratio of 1:100 salmon and 1:50 in 1993 (Fig. 14.4). There were less than one mobile *L. salmonis* per fish in these years but numbers of *Caligus* increased from July and in 1993 a treatment with dichlorvos was required and it appears that wrasse were ineffective in removing this species. However, with rapid and constant recruitment of *Caligus* from wild fish, it is difficult to confirm this. Cleaning of *Caligus* was reported by Kerrera Fisheries on Mull in 1990 and later years (Tomison, *pers. comm.*) where, following the introduction of wrasse at a ratio of 1:100 salmon, *Caligus* numbers declined each week over a four-week period by roughly one third each week. In a trial by McConnell Salmon at Loch Sunart in 1991, wrasse were stocked in July at a ratio of 1:100 (Mitchell, *pers. comm.*). Numbers of *Caligus* were low initially, followed by a gradual increase (Table 14.1). Wrasse were stocked in one group (B) but not in another (A). *Caligus* increased in both sampled cages at first, but declined to low levels in the cages with wrasse.

14.2   How widespread is the use of wrasse in Scotland?

The uptake in the use of wrasse on farms in Scotland and on the experiences/impressions of the technique by farmers was ascertained by circulating a question-naire. A list of companies was obtained from the Highlands and Islands Enterprise Directory of farmers and only companies on the Scottish mainland and Western Isles were circulated (a total of 56 companies). Several of these companies

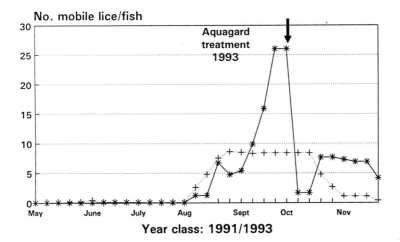

Fig. 14.4 Mean number of *Caligus elongatus* per post-smolt (*n* = 30 fish, each data point) when wrasse were stocked in 1991 (cross) and 1993 (star). The date of treatment with dichlorvos in 1993 is shown.

Table 14.1 *Caligus elongatus* numbers on salmon maintained in cages with and without wrasse.

Date	Mean number of *Caligus* per fish	
	Control (no wrasse)	Wrasse 1:100
26 August 1991	9.0	12.0
3 September 1991	22.0	6.0
9 September 1991	27.5	1.0

operated only one farm while some had several. As only 12 (21.4%) completed questionnaires were returned, further information was collected by telephone.

All farms had heard of the method and it appears to have been widely publicised in the popular and fish farming press. However, 77% of companies had not used wrasse (representing 66% of farms). The reasons given for not using wrasse were:

(1)   sea lice had not been a particular problem on that farm;
(2)   problems in obtaining a supply of wrasse;
(3)   concern regarding importing infectious agents with wrasse; and
(4)   cost was too high.

Of the 13 companies that had used wrasse, only one had found their use un-favourable. This company admitted that the husbandry of wrasse had not been as good as it had been in previous years.

14.2.1   *Examples of bad wrasse husbandry*

(1)   Lack of enough, and properly-designed, hides.
(2)   Insufficient care taken when changing nets.
(3)   Insufficient care during removal of dead salmon from the cages.
(4)   Nets frequently fouled.
(5)   Furunculosis had been an underlying threat through most of the production cycle and wrasse had not been vaccinated.
(6)   Numbers of lice had been high and the remaining wrasse had problems in coping with lice numbers.

14.2.2   *Favourable comments*

(1)   Wrasse effective in the first year of the production cycle.
(2)   20–30% reduction in lice numbers compared with previous production cycles.
(3)   No chemical treatments required in the first year of cycle while using wrasse (5 of 13 companies).
(4)   An overall reduction in mobile *L. salmonis* from 10 to 1 lice.

The positive respondents stated that a total of 150 000 wrasse had been stocked in 1994, and at a ratio of 1:50 salmon, this was equivalent to 39% of the total number of smolts put to sea in Scotland being cleaned by wrasse.

14.2.3   *Unfavourable comments*

(1)   Retention of wrasse over winter was poor with few fish remaining by April (four companies).
(2)   Ineffective against *Caligus*.
(3)   Net fouling had to be reduced to achieve efficacy.
(4)   Wrasse prone to cormorant attack.
(5)   Salmon over 2 kg ate wrasse (comment by two of 13 farmers).

14.3   Validation of the technique: evidence from Scotland

As far as is known, no laboratory or tank trials have been conducted in Scotland to determine the cleaning ability of wrasse. However, certain farmers have placed smaller salmon with high lice numbers in small cages with wrasse to allow fish to become trained in eating lice, and lice numbers on these salmon have declined quickly (Tomison, *pers. comm.*). Farm staff have observed wrasse from the surface picking lice from salmon (Smith, *pers. comm.*). Additionally, by using an underwater camera, wrasse were observed to swim quickly upward into a school of salmon and to pick lice from them (Treasurer, 1993).

An attempt was made to use a small stomach pump to extract the gut contents of anaesthetized wrasse. However, the bore of pipette required was too small to permit recovery of gut contents. A small number of wrasse was sacrificed and gut

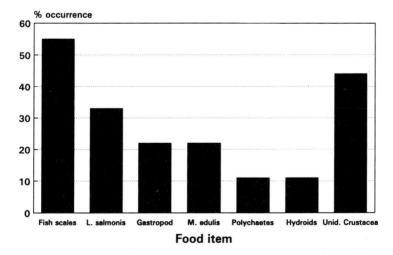

Fig. 14.5 Percentage occurrence of food items in the alimentary tracts of nine goldsinny wrasse of 116–27 mm total length.

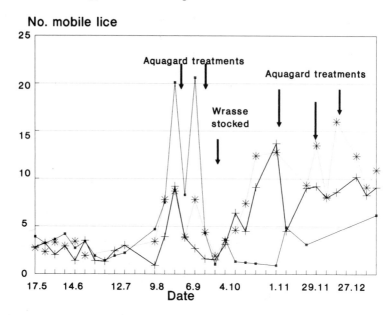

Fig. 14.6 A comparison of numbers of mobile *L. salmonis* on salmon post-smolts on one group of cages stocked with wrasse (squares) with two groups with no wrasse (cross and stars) on a farm in Mull in 1993. Wrasse were stocked with salmon in cages only on group 1 on 28 August. Arrows indicate treatments with dichlorvos. Fish on group 1 were not treated with dichlorvos after August.

contents examined (Fig. 14.5). Five of nine fish were seen to have consumed lice although a variety of organisms was found in the diet.

Another method of validating the efficacy of wrasse is to compare lice numbers in cages with and without wrasse, and various examples have been given above. A recent case is from a farm on Mull (Fig. 14.6). The farm comprised three groups of ten cages, eight on each group held fish with *c.* 100 000 salmon of the 1993 year class. After stocking in April, *Caligus* numbers increased first, followed by an *L. salmonis* increase from July. Lice numbers in group 1 (20 mobile lice per fish) were consistently higher than in groups 2 and 3 (7–8 mobile lice). Wrasse (a mixture of goldsinny and rock cook) were stocked on 20 August only in cages in group 1 at a ratio of 1:50 salmon. Thereafter, lice numbers in group 1 remained below 10 per fish and only increased from December with the onset of colder temperatures. In groups 2 and 3 lice numbers rose to 15 per fish and three further treatments with dichlorvos were required by December.

14.4   Practices in the use of wrasse to control lice on salmon farms in Scotland

From experiences of using wrasse on a commercial scale the following guidelines/ practices are followed by salmon farmers in the main Scottish companies:

(1) On receipt from fishermen, wrasse are checked to ensure that only goldsinny, rock cook and corkwing are accepted. Ballan can be aggressive to salmon (Costello & Bjordal, 1990) and cuckoo wrasse are relatively rare in inshore waters of west Scotland. The minimum length of wrasse to be accepted should be 100 mm (Chapter 7, this volume), as smaller fish may escape or become gilled in the netting. The condition of wrasse should be examined for erosion of fins, to ensure that wrasse have been held in adequate conditions.

(2) A sample of wrasse should be screened for potential infectious pathogens if fish are coming from a new source. Wrasse should be vaccinated with a suitable vaccine to protect against *Aeromonas salmonicida* (this volume, Chapter 20; Treasurer & Laidler, 1994).

(3) Wrasse may be held in a holding cage to permit domestication prior to stocking in salmon cages and to give sufficient time for health screening. It may be helpful to train wrasse by introduction of louse-infected salmon to this holding cage. There is a high risk of wrasse being the subject of bird predation in a holding cage and adequate hides and protection should therefore be provided.

(4) The recommended stocking ratio is 1 wrasse : 50 salmon.

(5) Wrasse should be fed during the holding period or when lice numbers are low, with shellfish such as crushed mussel *Mytilus edulis* L., scattered in cages or held in muslin bags.

(6) On stocking, hides should be placed in the cages. There are many designs, e.g. plastic pipes, tyres tied together. These hides should be suspended low in the net but about 1–2 m above the bottom of the net. Hides provide refuge or suitable habitat within the salmon cage, thus reducing stress and enhancing survival of wrasse. They also keep the wrasse from using the mortality collecting sock as a hide, protect against bird predation and against strong currents. Wrasse are weak swimmers (Hilldén, 1981) and it is likely that they are highly stressed by maintaining position in strong currents. This may lead to mortality, in part increased by the effects of low temperature or reduced salinity (this volume, Chapter 10). Hides can also be used for transfer of wrasse between cages and for overwintering, as provision of adequate and sufficient hides greatly enhances survival (this volume, Chapter 22).

(7) The mortality sock should be tied when wrasse are introduced to the cages and for a period of days to encourage wrasse to commence cleaning activity.

(8) Nets should not be too fouled as this reduces cleaning activity by wrasse as they graze on the net (Fig. 14.7).

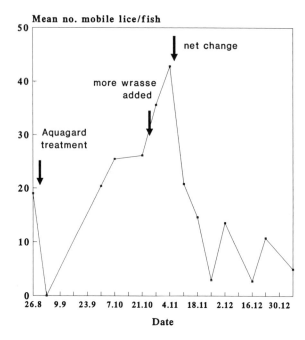

Fig. 14.7   The effects of a net change on cleaning activity of goldsinny on a farm in Loch Eil, 1991. The addition of more wrasse did not reduce lice numbers but lice numbers declined following a net change, suggesting that wrasse had previously grazed on the net. Ten salmon were sampled on each date.

(9)   Nets should be routinely checked for holes.

(10)   Wrasse should not be used on farms exposed to low salinity (<20‰) for more than 24 hours. There may be a surface layer of reduced salinity and a net raised for routine farm operations should be lowered as soon as possible.

(11)   Care should be taken to ensure that wrasse are not enmeshed when changing nets.

(12)   If mortalities of wrasse are detected the cause should be determined and, if due to an infectious agent, wrasse should be removed by introducing traps. Consideration should be given to treatment by injection with an antibiotic (Treasurer & Laidler, 1994).

(13)   If lice numbers reach a level requiring treatment, despite wrasse being stocked, chemical control measures should be used. There is evidence that wrasse survive sea lice treatments with organophosphates such as dichlorvos (Treasurer, 1991), azamethiphos (Roth, *unpubl. rep.*) and hydrogen peroxide (Bruno & Raynard, 1994).

(14)   Wrasse should be transferred between cages if lice are a greater problem in other cages.

14.5 Conclusions

The use of wrasse as cleaner-fish of sea lice is now well established on suitable salmon farms in Scotland. From a questionnaire sent to salmon farmers in Scotland it was revealed that wrasse were stocked with a significant proportion (39%) of the smolt input in 1994. Several examples have been given of control of *L. salmonis* in the first year of the salmon production cycle. However, wrasse are lost from cages over winter and fish farmers suggested this problem should be examined further, particularly with reference to the improvement in the design of wrasse refuges and to the effect of environmental conditions on survival.

Acknowledgements

I am grateful to the respondents to the questionnaire for their comments and to the many staff involved in the development of the use of wrasse on salmon farms.

References

Bjordal, A. (1991) Wrasse as cleaner-fish for farmed salmon. *Progress in Underwater Science* **16**: 17–29.

Bruno, D. and Raynard, R. (1994) Studies on the use of hydrogen peroxide as a method for the control of sea lice on Atlantic salmon. *Aquaculture International* **2**: 10–18.

Costello, M. and Bjordal, A. (1990) How good is this natural control on sea lice? *Fish Farmer* **13**: 44–6.

Treasurer, J.W. (1991) Limitations in the use of wrasse. *Fish Farmer* **14(5)**: 12–13.

Treasurer, J.W. (1993) Management of sea lice (Caligidae) with wrasse (Labridae) on Atlantic salmon, *Salmo salar* L., farms. In *Pathogens of Wild and Farmed Fish: Sea Lice* (Ed. by Boxshall, G.A. and Defaye, D.). Ellis Horwood Ltd., Chichester, pp. 335–45.

Treasurer, J.W. (1994) Prey selection and daily food consumption by a cleaner fish, *Ctenolabrus rupestris* (L.), on farmed Atlantic salmon, *Salmo salar* L. *Aquaculture* **122**: 269–77.

Treasurer, J.W. and Laidler, L.A. (1994) *Aeromonas salmonicida* infection in wrasse (Labridae), used as cleaner-fish, on an Atlantic salmon, *Salmo salar* L., farm. *Journal of Fish Diseases* **17**: 155–61.

Chapter 15

Large-scale use of wrasse to control sea lice and net fouling in salmon farms in Norway

P.G. KVENSETH *A/S MOWI, Bontelabo 2, P.O. Box 4102, Dreggen, N–5023 Bergen, Norway*

Wrasse (Labridae) have been used by A/S MOWI since 1989 in increasing numbers as cleaner-fish to remove sea lice (*Lepeophtheirus salmonis* Krøyer and *Caligus elongatus* L.) from farmed Atlantic salmon (*Salmo salar* L.) in western Norway. At one site, Skorpeosen west of Bergen, wrasse (*c.* 106 000 in total) were stocked with one salmon generation from smolt release in May 1992 until harvesting 18 months later, and the degree of sea lice infestation quantified.

The efficiency of wrasse as cleaners was good for salmon smaller than 2 kg during the first year at sea. Few wrasse survived over the winter when stocked together with salmon in netpens. The salmon reached an average weight of 3–6 kg during the second year at sea (1993), and several problems with the use of wrasse were experienced. Gut analysis revealed 50–100% with no food items, but a maximum of 32 lice were found in one goldsinny [*Ctenolabrus rupestris* (L)]. Other food items were fragments of blue mussel (*Mytilus edulis* L.), amphipods and algal fragments. The number of wrasse declined rapidly after each addition to the pens because of damage from handling, bacterial diseases and possible predation from salmon.

Wrasse were also effective in reducing the extent of net fouling. At A/S MOWI's marine site at Åkre in Hardanger, wrasse added in May 1993 reduced the number of net changings by 50%, compared with control nets without wrasse. It is concluded that cleaner-fish represent an economically beneficial way of reducing the problems with salmon lice infestations and net-fouling and assist in reducing the use of organophosphate pesticides and copper containing net paints.

15.1 Introduction

Wrasse are used as cleaner-fish to remove sea lice from farmed Atlantic salmon (*Salmo salar* L.) in Norway (Bjordal, 1991; Kvenseth, 1993), Ireland (Darwall *et al.*, 1991) and Scotland (Treasurer, 1993). The cleaning effect of using wrasse has been documented for post-smolt and salmon up to 2 kg (Treasurer, 1993). However, wrasse have been less effective in cleaning salmon larger than 2 kg and their inclusion with salmon during the second year at sea has been sparse, and with varying effect (Chapter 13, this volume). The proven financial and environmental benefits of using wrasse would make their use to control sea lice infestations for the entire period that the salmon are at sea desirable. A/S MOWI started experiments in 1992 to assess the possibility of only using wrasse on sites where chemical treatments had previously been necessary.

Earlier work (Hildén, 1978; Bjordal, 1991; Costello & Donnelly, 1991), and observations made on MOWI farms observed wrasse grazing on the surface of seaweed and nets. An additional aim of this study was therefore to test if this feeding behaviour was strong enough to reduce the number of net changes and consequently reduce the use of antifouling paints.

15.2   Materials and methods

15.2.1   *Experimental site*

The salmon farm at Skorpeosen was established as a new location in April 1992 (six Aqualine net pens). In May, 530 000 salmon smolts, at an average weight of 70 g, were stocked. Smolts were vaccinated against *Aeromonas salmonicida* infections before transfer and revaccinated by injections during February–March 1993. The net pens were 70 m in circumference, mounted with 10 m deep nets of 12 × 12 mm or 16 × 16 mm square mesh. As the salmon grew, additional pens were stocked, and the salmon biomass was redistributed to a total of 16 pens by April 1993. Salmon growth was monitored monthly by weighing 100 salmon at random from each pen. A total of 1800 tonnes of salmon was harvested from June to December 1993. Sea temperature was monitored daily at 1 m and 6 m.

15.2.2   *Wrasse*

Wrasse were caught locally by fishermen, with goldsinny [*Ctenolabrus rupestris* (L.) as the dominant species (90%). The remaining 10% were a mixture of rock cook [*Centrolabrus exoletus* (L.)] and corkwing [*Crenilabrus melops* (L.)]. Wrasse were added to all pens, reaching a recommended density of 1:50 post-smolt (Costello & Bjordal, 1990) during late June 1992 (Fig. 15.1). Total additions of wrasse in 1992 were 16 000, with dead wrasse replaced until September 1992. More wrasse were stocked in May 1993, with total additions of 90 000 by October 1993. Several shelters were provided in all pens for the cleaner-fish. Dead and moribund wrasse were screened for bacterial diseases during July and August 1993.

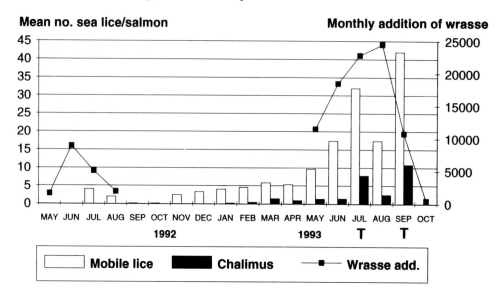

Fig. 15.1 Average infections of sea lice at Skorpeosen 1992–93. Total number of salmon examined = 5967; total addition of wrasse = 106 000; T = treatment with hydrogen peroxide.

15.2.3   *Lice counting*

Salmon lice were counted at one- to three-week intervals, more frequently in summer than winter months. At least 20 salmon were sampled at random from each pen by a dipnet lowered into the cage, and anaesthetized by MS 222 or benzocaine. Mobile lice were counted, and the salmon were returned to the pen. In 1993 chalimus stages were also recorded.

15.2.4   *Gut content analysis of wrasse*

The gut contents of wrasse were examined sporadically during 1992. A more organized examination was conducted in one pen during the period May–October 1993, where, at weekly intervals, 20 wrasse were killed by prolonged anaesthesia in benzocaine and the entire alimentary tract opened (289 fish were examined in total). Wrasse were sampled from wrasse refuges hauled to the surface. Four gut content categories were identified: sea lice (*Lepeophtheirus salmonis* Krøyer and *Caligus elongatus* Nordmann), blue mussel (*Mytilus edulis* L.), amphipods and algae. Where no items were present, the gut was classified as empty.

15.2.5   *Net fouling*

At a salmon farm located in a fjord at Åkre in Hardanger, large-scale blue mussel settlement and rapid growth resulted in excessive net fouling. There were however no problems with sea lice. The pens at Åkre, mounted with 12 m deep nets were impregnated with standard copper containing anti-fouling paint. Simultaneously with the release of smolts in May 1993, eight pens were each stocked with 500 goldsinny, while four pens were kept without wrasse as controls. Two of the control pens were stocked with 500 goldsinny during June 1993.

15.3   Results and discussion

15.3.1   *Sea temperature*

The sea temperature at Skorpeosen showed only minor differences between 1 m and 6 m. The temperature increased from 7°C in May 1992 to 14–16°C three weeks later, and remained stable until October, followed by a steady decrease to 6–7°C in late December. The minimum sea temperature during the winter was 4.8°C in March. During April and May 1993 the temperature increased to 10°C, and stabilized between 10°C and 14°C until November.

15.3.2   *Salmon growth*

The weight of salmon is given as an average for each month, with data pooled for all pens (Fig. 15.2). Average weight reached 1 kg by November 1992, 2 kg by May 1993 and 5 kg during November–December 1993.

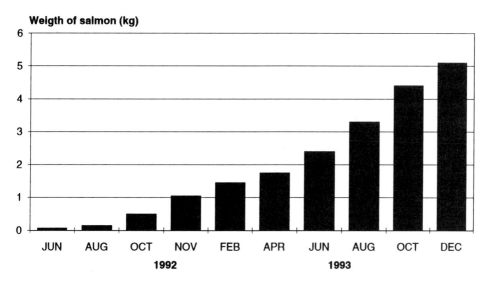

Fig. 15.2   Growth of salmon at Skorpeosen 1992–93.

15.3.4   *Wrasse*

Problems with controlling the stocking density of wrasse were demonstrated when all pens were emptied and the salmon redistributed during February and March of 1993. The calculated number of wrasse should have been 10 000 based on the numbers added and the dead wrasse removed, but less than 100 were found. Large wrasse losses were also experienced in late 1993 as 35 000 were added during July and August, but only a few hundred were found during the harvesting period (September–December). Some of the wrasse introduced were too small (less than 10 cm) and possibly escaped through the net meshes; some were gilled in the net. Dead wrasse disintegrated quickly at summer temperatures and the remains may have been lost through the net mesh, resulting in underestimation of mortality levels.

15.3.5   *Bacterial diseases*

During the spring and summer of 1993 high mortality was recorded among added wrasse. Bacteriological screenings of moribund fish were dominated by atypical *A. salmonicida*. Atypical *A. salmonicida* is found latent in several wild stock of fish: cod (*Gadus morhua* L.) (Cornick *et al.*, 1984); sandeels [*Ammodytes tobianus* L. and *Hyperoplus lanceolatus* (Le Sauvage); Dalsgaard & Paulsen, 1986] and goldsinny wrasse (Frerichs *et al.*, 1992). Stressing wrasse which carry a latent infection, through fishing, handling and transportation, may lead to an outbreak of bacterial diseases, resulting in extensive mortalities. Wrasse must therefore be handled with care through catching, transportation and treatment in cages, to keep them in a healthy condition. Isolates of atypical *A. salmonicida* from wrasse have been found non-pathogenic to experimentally infected salmon smolt (Frerichs *et al.*, 1992), and salmon infected with atypical *A. salmonicida* were not detected during these experiments.

15.3.6   *Sea lice infection*

C. elongatus was the dominant lice species during June 1992 at Skorpeosen, with an average initial density of approximately four per post-smolt during July (Fig. 15.1). This value decreased during the summer to an average of 0.1 lice per fish, and remained at this level until November. *L. salmonis* infected the salmon during November 1992, and were thereafter the dominant species, with an average density of about five mobile lice per salmon through the winter. As the temperature increased in May 1993, the mean number of mobile lice increased to 10 per salmon, and further to about 20 in June and 30 in late July (Fig. 15.1). At this time the lice infection had reached a level probably above the limits of control for the wrasse.

All pens were treated with hydrogen peroxide in July 1993. In spite of

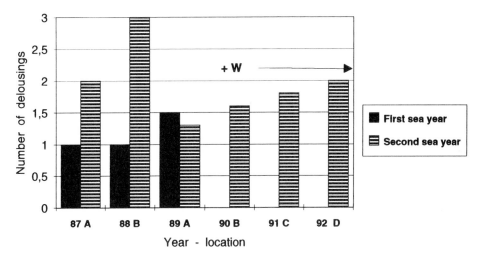

Fig. 15.3 Delousing frequencies for different salmon generations and different locations (A–D) in the same area run by MOWI. 92D = Skorpeosen; +W = wrasse added at smolt release.

the successful treatment and further addition of 35 000 wrasse in August and September, the lice infection increased to 40 mobile lice per salmon, and another treatment with hydrogen peroxide was necessary in late September (Fig. 15.1). As the direct comparison of pens with and without wrasse at Skorpeosen was not possible, the situation in 1992/93 was compared with earlier generations of salmon in the same area (Fig. 15.3). For the 1987, 1988 and 1989 generations, one to two 'Nuvan' treatments were necessary during the first eight months at sea (Bjordal, 1991), and one to three additional treatments during the second sea-year. For the 1990 and 1991 generations, wrasse were added at smolt release and no delousing was necessary. This compares with no delousing in the 1992 season with salmon <2 kg and two treatments with hydrogen peroxide on bigger salmon in 1993 (Fig. 15.3). As problems with salmon lice infestations increased from 1989 for other salmon farms in the immediate area, the decreases recorded in the farms stocked with wrasse cannot be explained by natural reductions in sea lice numbers.

15.3.7 *Wrasse gut contents*

From May to July 1993 sea lice were found in 10–15% of guts while the rest of opened wrasse were empty of all food items (Fig. 15.4). In August and September most of the wrasse had eaten algae, amphipods and mussels, with a low prevalence of lice (5–10%) (Fig. 15.4). This possibly indicates that wrasse find organisms living on semi–stable surfaces (net) easier to catch than lice on swimming salmon, even if lice are available in increasing numbers. Therefore in order to ensure wrasse are actively grazing sea lice, net fouling should be kept at a low level to reduce access to alternative, potentially more available, food items. Another

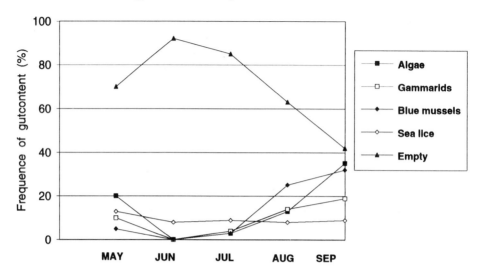

Fig. 15.4   Average frequency of gut contents. Percentage of wrasse found with different food items or empty.

option would be to increase the density of wrasse as fouling increases, to maintain a sufficient control over both lice infection and net fouling.

15.3.8   *Limitations of wrasse use*

The reasons why wrasse could not keep sea lice under control through the second summer at sea may be explained by high wrasse mortality from bacterial diseases, in combination with escapes and gilled wrasse. This made estimating and maintaining an appropriate density of wrasse difficult. In addition, large salmon (approximately 2 kg) were observed to attack and eat wrasse. Although losses were difficult to estimate, increased lice infestation and net fouling coincided with the most severe reductions in wrasse numbers. At this time, alternative food sources for the wrasse, such as amphipods, algae and blue mussels living on the surface of fouled nets, were easily accessible. During the second summer at sea the salmon had increased tenfold in body weight compared with post-smolts. As the use of fish feed had risen concomitantly, increased nutrients were probably added to the surrounding water from feed and salmon faeces, possibly enhancing algae growth and net foulings.

15.3.9   *Net fouling*

By June 1993, one month after initial stocking with wrasse, clear differences were observed between pens with and without wrasse. Blue mussel spats had settled on pens without wrasse, while pens with wrasse remained clean. During June, wrasse were added to two of the control pens. In August the two remaining control pens were overgrown, mainly with blue mussels, and the nets had to be changed. Nets in the pens with wrasse added, either in May or in June, stayed clean through August and were changed late in autumn, during salmon redistribution to other pens.

Acknowledgements

I am grateful to Jan K. Nielsen for examining gut contents and to staff at Skorpeosen and Åkre fish farms for helpful assistance during the experiments. I am also grateful to Anne Mette Kvenseth for suggesting improvements to the manuscript. This study was supported financially by NFR, the Norwegian Research Council.

References

Bjordal, Å. (1991) Wrasse as cleaner-fish for farmed salmon. *Progress in Underwater Science* **16**: 17–28.

Cornick, J.W., Morrison, C.M., Zwicker, B. and Shum, G. (1984) Atypical *Aeromonas salmonicida* infection in Atlantic cod, *Gadus morhua* L. *Journal of Fish Diseases* **7**: 495–99.

Costello, M.J. and Bjordal, Å. (1990) How good is this natural control of sea-lice? *Fish Farmer* **13(3)**: 44–6.

Costello, M.J. and Donnelly, R. (1991) Development of Wrasse Technology. In *Proceedings from Irish Salmon Growers Association 5th Annual Conference and Trade Exhibition* (Ed. by Joyce, J.R.). Irish Salmon Growers Association, Dublin, pp. 18–20.

Dalgaard, I. and Paulsen, H. (1986) Atypical *Aeromonas salmonicida* isolated from diseased sand-eels, *Ammodytes lancea* (Curvier) and *Hyperoplus lanceolatus* (Le Sauvege). *Journal of Fish Diseases* **9**: 361–4.

Darwall, W., Costello, M.J. and Lysaght, S. (1992) Wrasse: how well do they work? *Aquaculture Ireland* **50**: 26–9.

Frerichs, G.N., Millar, S.D. and McManus, C. (1992) Atypical *Aeromonas salmonicida* isolated from healthy wrasse (*Ctenolabrus rupestris*). *Bulletin of the European Association of Fish Pathologists* **12**: 48–9.

Hilldén, N.O. (1978) On the feeding of the goldsinny, *Ctenolabrus rupestris* L. *Ophelia* **17**: 195–8.

Kvenseth, P.G. (1993) Use of wrasse to control salmon lice. In *Fish Farming Technology* (Ed. by Reinertsen, H., Dahle, L.A., Jørgensen, L. and Tvinnereim, K.). Balkema, Rotterdam, pp. 227–32.

Treasurer, J.W. (1993) Management of sea lice (Caligidae) with wrasse (Labridae) on Atlantic salmon (*Salmo salar* L.) farms. In *Pathogens of Wild and Farmed Fish: Sea Lice* (Ed. by Boxshall, G.A. and Defaye, D.). Ellis Harwood, London, pp. 335–45.

Treasurer, J.W. and Laidler, L.A. (1994) *Aeromonas salmonicida* infection in wrasse (Labridae) used as cleaner fish, on Atlantic salmon, *Salmo salar* L., farm. *Journal of Fish Diseases* **17**: 155–61.

Chapter 16
Wrasse as cleaner-fish: the Shetland experience

C.M. YOUNG *Vaila Sound Salmon Ltd, Backlands, Walls, Shetland ZE2 9PD, UK*

Trials using wrasse as the primary treatment for sea lice infestation, commenced in Shetland salmon farms in 1989. The effectiveness of using wrasse demonstrated in intial trials resulted in wrasse being introduced throughout Shetland farms on an experimental basis. Results from this more extensive trial were more variable depending on the conditions within respective farms. One of the major drawbacks for the use of wrasse in Shetland was the relative sparsity of local populations necessitating importation. Nevertheless, wrasse continue to be used by some farms in Shetland. Recommendations for the use of wrasse are given.

16.1   Initial trials

In 1989, aware of reports from Norway on the work of Åsmund Bjordal in using wrasse to control sea lice on salmon farms, the Shetland Salmon Farmers' Association (SSFA), the Sea Fish Industry Authority at Ardtoe and the Fiskeri-forskning-sinstitut in Bergen were approached by the author for information. Ardtoe was looking for financial support to investigate the application of wrasse and, with help from the Shetland Islands Council, the SSFA offered funding. At the time there was a concerted media attack on the industry regarding use of dichlorvos, and alternative treatments were worth investigating.

The initial trial was intended to demonstrate whether a commercial application would work in total rather than only, for example, a 10% decrease in lice numbers. The first trial took place in October 1989, and used 100 goldsinny, *Ctenolabrus rupestris* (L.), caught near Ardtoe, west Scotland, and flown to Shetland. The trial was conducted in a block of 6 × 12 m cages; all cages were treated prior to the trial. The wrasse were introduced into one cage of 5200 salmon post-smolts of *c*. 400 g; the other five cages received no further treatment. On this site there was a significant sea lice problem that had required treatment every month during that summer. One month later the five cages without wrasse needed treatment whereas the one with wrasse contained clean salmon showing increased vigour and obvious improved health (Table 16.1). The experiment was continued by treating four of the five remaining infested cages with dichlorvos before transferring 67 wrasse from the initial trial cage to the fifth cage previously

Table 16.1 Initial trial of wrasse cleaning at Vaila Sound Salmon Ltd, Shetland, October 1989 (reproduced with permission from Smith, 1990).

Date	5 Control cages (no wrasse)	Trial cage (wrasse)
23 September 1989	Dichlorvos	Dichlorvos
28 September 1989	–	100 wrasse added
3 October 1989	10 sea lice/fish some damage	10 sea lice/fish some damage
19 October 1989	10 sea lice/fish damage	zero sea lice/fish damage
31 October 1989	20 sea lice/fish damage	zero lice, zero damage

Table 16.2 Second trial of wrasse cleaning at Vaila Sound Salmon Ltd, Shetland, October 1989 (reproduced with permission from Smith, 1990).

Date	4 Control cages (no wrasse)	Original trial cage	New trial cage
6 November 1989	dichlorvos	–	–
7 November 1989	zero sea lice	zero sea lice	20 lice/fish
9 November 1989	–	33 wrasse left in	67 wrasse added
12 November 1989	zero lice zero damage	zero lice zero damage	5 lice/fish damage
19 November 1989	zero damage	zero damage	damage
27 November 1989	5 lice/fish damage start	zero lice zero damage	3 lice/fish less damage
19 December 1989	25 lice/fish	<1 louse/fish	3 lice/fish

without wrasse. Cleaning was again demonstrated with remarkable success (Table 16.2) and the potential for lice control led to further trials in Shetland.

16.2   Trials throughout Shetland

It became clear that unlike Norway and the west of Scotland, Shetland had few native wrasse. Initially, salmon farmers fished in their locality, but without success. Divers confirmed that the Gonfirth area was the only site with sufficient wrasse to justify fishing (Costello *et al.*, 1993). Although the habitat appeared suitable, temperature and other conditions may have been inadequate to support a viable breeding population (Darwall, *pers. comm.*). Therefore, in 1990, the SSFA sponsored the shipment and distribution of trial quantities of wrasse to interested farmers.

As supplies had to be imported from the west of Scotland, cost was a major factor. The biomass was too small to justify a wellboat and whatever method was used (even a plane was chartered direct from Mull to Shetland), transport to Shetland was so costly that freight added about £1.50 per fish to the existing cost of £2 each. The first shipment were mixed goldsinny and rock cook imported from the west of Scotland by air and quarantined at an offshore site for six weeks. Wrasse were then distributed by the SSFA at cost on 7 September 1990. A second consignment, by road, sea and road again, arrived on 30 October 1990. The surface transportation (two days) and some subsidy from the SSFA had cut the cost to £2.50 per fish. Fifteen farms (a quarter of the Shetland farms) stocked at least one cage with wrasse, a mix of approximately 50:50 goldsinny and rock cook, *Centrolabrus exoletus* (L.). A survey was commissioned by the SSFA and carried out by Leslie Copland in conjunction with Heriot-Watt University in 1991. The results varied from total success to failure (Table 16.3).

Farmers were interviewed and details taken of cage-type, salmon:wrasse ratio, and hide-use. Farmers who did not take wrasse were also interviewed as to their reasons. Other farmers found no difference as a result of taking wrasse (Table 16.3), perhaps from reluctance to trust their stock to such a radical technique. Some peculiar results became apparent – the most bizarre being that wrasse worked better in square as compared with round cages. This is presumably because of the space in the corners of the rectangular net left by salmon swimming in a shoal. Other factors involved were the wrasse:salmon ratio (1:50 seemed optimum), provision of hides to reduce stress; cleanliness of nets to minimize populations of crustacean prey; and full strength salinity. Mesh size of 12 mm or less was critical as wrasse escaped from larger mesh sizes. This was incompatible with the salmon farmers' practice of using larger mesh nets as the stock increase in size.

It was clear that under certain circumstances, sea lice infestation was controlled even under conditions of heavy challenge. In one case, wrasse were stocked into one cage of twelve. That cage only required a treatment after five months (and after a reduction in the wrasse/salmon ratio) whereas the other eleven cages were treated approximately every three weeks (Copland, 1991).

Loss of wrasse was a problem. In the commercial environment, many wrasse disappeared without trace. Some may have escaped through small holes; some may have been taken by birds above and below the water. However, one farm succeeded in catching goldsinny locally and reported that 80% survived beyond a six-month period, in contrast to the farms with imported wrasse, where only two had more than 50% of wrasse remaining after six months. Some reported total losses in this period (Copland, 1991). It would seem that the stress associated with holding and transportation, as well as the salmon cage environment, may be significant mortality factors.

Those farms in the trial that were considered most successful provided shelters in the form of short lengths of pipe. The difference was not striking due to the relatively small numbers of shelters provided, with an average of eight per cage (Copland, 1991). Perhaps more shelters per cage may be more beneficial.

The foremost reason for farms not participating in the trial, except for having no lice problem, was the fear of introducing disease (Table 16.4). Other reasons were disbelief in such a radical proposal, cost and difficulties in using 12 mm mesh

Table 16.3 Results of 1990/1991 trials with wrasse in Shetland (from Copland, 1991).

Farm health (1 = best, 13 = worst)	Introduction[1]	Estimate of losses	Shelter	Stocking ratio (wrasse:salmon)	Cage shape	Salmon size at introduction (g)	Salmon stocking density (kg m^{-3})	Improved growth
1	continuous	17%	yes	1:50	square	150	2.5	yes
2	1 and 2	<half	yes	1:50	square and round	300–500	1.5–3.0	yes
3	1	most	yes	1:50	12-sided	1000	10.0	yes
4	1 and 2	most	yes	1:50 and 1:100	round	250–500	2.5–5.0	yes
5	2	half	yes	1:83	round	800	10.0	no
6	1	all	no	1:100 and 1:200	square	100	1.0–2.0	no
7	2	most	no	1:100	square	700	10.0	yes
8	2	all	no	1:50	square	1000	10.0	no
9	1	all	no	1:120 and 1:100	round	600–800	7.0–9.0	no
10	2	all	no	1:150	round	800+	7.0	no
11	1	all	yes	1:33	round	650–700	4.5	no
12	2	most	yes	1:40	square	150–200	1.5	no
13	1	most	no	1:115	round	70	1.6	no

[1] Wrasse introduced in two stages, except where continuous.

Table 16.4   reasons indicated for non-purchase of wrasse by members of the SSFA in 1990 (from Copland, 1991).

Reasons for non-purchase	Given by SSFA members
Lice problem not severe	9
Worried about disease introduction	8
Unconvinced of their worth	7
Too expensive	4
Don't want to use 12 mm nets	4
Don't have enough 12 mm nets	3
Disease transfer between generations	1
Insufficient supply of wrasse	1
Caught own wrasse locally	1
Satisfied with dichlorvos	1

(Table 16.4). Only one respondent indicated that he was satisfied with Aquaguard as a control measure (Copland, 1991).

One Shetland farm, Gonfirth Salmon Ltd, succeeded in catching goldsinny and had considerable success in lice control. This is the ideal situation as the costs of wrasse supply are low (i.e. limited to the labour costs of setting and lifting creels), and any disease risk is eliminated. Acclimatization problems are also minimized and wrasse stocks can be constantly replenished when necessary. A number of other Shetland salmon farmers (including the author) tried fishing in their locale but without success, although ballan, *Labrus bergylta* Ascanius, and cuckoo, *Labrus mixtus* L., were present.

In 1992, as part of an EC FAR project, a team of divers investigated the Shetland marine habitat for wrasse populations and confirmed that the Gonfirth area was the only site where sufficient numbers of wrasse (goldsinny) justified fishing (Costello *et al.*, 1993). Although there were many areas of apparently suitable habitat, the marine season (temperature and photoperiod) was not as advanced as either Ireland or Norway and the conditions were less than ideal for a viable breeding population (Darwall, *pers. comm.*).

16.3   Drawbacks in using wrasse

Considerable practical experience has been gained since 1989, mostly through trial and error. Although the first trial carried out at Vaila Sound Salmon Ltd, showed remarkable results, later experience at this farm and others showed that the technique was not so simple and that there are a number of drawbacks.

In Shetland, the first difficulty was obtaining a supply of wrasse both early enough in the season and cheap enough to be practical. Wrasse need perhaps a month to acclimatize and start feeding on lice. Introduction of wrasse much later than the end of July with smolts stocked in April and May risks a build up of lice that is too much for the wrasse to cope with. Also, if wrasse are introduced to salmon that are heavier than about 800 g, the salmon predate on them and cleaning behaviour may not become established.

Perhaps the biggest drawback for the farmer is the requirement for 12 mm smolt nets to be used throughout the entire salmon grow-out cycle. It is customary to provide the larger salmon with nets of 18 mm, 22 mm or larger mesh to promote free-flow of clean well-oxygenated water. Nets of 12 mm foul more quickly and the faeces of >2 kg salmon do not fall readily through the net bottom and have to be pumped out at regular intervals. However, the faster-growing corkwing, *Crenilabrus melops* (L.) can be kept in mesh larger than 12 mm.

In the Shetland trials, wrasse successfully cleaned salmon to a weight of over 4 kg if introduction was early enough in the cycle. It seems that the two species become conditioned to the relationship if initiated when there is not too much size difference. Restocking in the second spring was also useful as the new intake of wrasse seem to fit in well with the remaining population and the wrasse to salmon ratio could be restored.

One useful technique was to harvest progressively and as each cage is emptied the wrasse are transferred to the cage that needs them most thus concentrating the remaining wrasse into the final cages. Ideally, wrasse are caught from the cage before the starvation period as the sharpened appetite overcomes the conditioned behaviour. As some of the preferred prey items of wrasse are crustaceans (this volume, Chapter 2), it is logical to assume (and experience has shown) that clean nets help to keep the wrasse cleaning salmon, as the enormous supply of amphipods living on a fouled net could provide ample feed without the wrasse having to take lice.

Shetland's need to import wrasse stocks from the west of Scotland raised the fear of importing disease. It has been demonstrated that goldsinny can carry, and die of, furunculosis (Treasurer & Cox, 1991). For this reason, some of the earlier stock were quarantined at a vacant offshore site for testing and observation prior to distribution, but no pathogens were identified. While there have been outbreaks of disease over the period that wrasse have been in use in Shetland, bacterial sensitivity tests to antibiotics indicated that the source was not related to wrasse imports.

Hatchery-reared wrasse are an obvious solution to any potential disease problem and in June 1993, 67 hatchery-bred goldsinny were introduced to a commercial cage at Vaila Sound Salmon Ltd, for comparison with wild caught fish. Prior to introduction, these wrasse were vaccinated against furunculosis by injection of Furogen 009. No discernable difference in cleaning ability was observed although mortality was higher than usual in the hatchery-bred fish.

16.4   Conclusions

Lice populations were kept consistently low using wrasse, compared with the wide cyclical fluctuation in levels experienced when using dichlorvos. However, experience with wrasse in Shetland has not persuaded the local industry to adopt them to a significant degree. Wrasse-use in Shetland has declined and the author's farm remains the only user at this time. The three main reasons for this are high transportation costs, the need for 12 mm nets and the threat of disease transmission. Local hatchery production would resolve the transportation problem and remove the threat of disease transmission. Perhaps the use of faster-growing corkwing (this volume, Chapter 12) could overcome the mesh-size problem.

In the absence of anything better, the use of wrasse for lice control is recommended, given the following conditions:

(a)   a local (or hatchery) supply of rock cook, goldsinny or corkwing;
(b)   1:50 wrasse to salmon ratio;
(c)   12 mm clean nets;
(d)   ample hides of pipes or other shelters;
(e)   early introduction to smolts; and
(f)   restocking in the second spring.

References

Copland, L. (1991) *A study of wrasse as cleaner fish for the control of sea lice in the salmon farming industry*. Unpublished MSc thesis, Institute of Offshore Engineering, Heriot-Watt University, Edinburgh.

Costello, M.J. *et al.* (1993) Cleaner-fish technology: parasite control, alternative to pesticides, and creating a new fishery. EC 'FAR' Project Report, Dublin.

Smith, P. (1990) Preliminary Scottish trials with wrasse for the control of sea lice. SFIA (*unpubl. rep.*).

Treasurer, J. and Cox, D. (1991) The occurrence of *Aeromonas salmonicida* in wrasse (Labridae) and implications for Atlantic salmon farming. *Bulletin of the European Association of Fish Pathologists* **11**: 208–10.

Chapter 17

Parasites and diseases of wrasse being used as cleaner-fish on salmon farms in Ireland and Scotland

M.J. COSTELLO[1], S. DEADY[2], A. PIKE[3] and J.M. FIVES[2]

[1] *Environmental Sciences Unit, Trinity College, University of Dublin, Dublin 2, Ireland;*
[2] *Department of Zoology, University College, Galway, Ireland; and* [3] *Department of Zoology, University of Aberdeen, Tillydrone Avenue, Aberdeen AB9 2TN, UK*

A literature review of the diseases and parasites of north European wrasse (Labridae) was conducted. Over 800 corkwing (*Crenilabrus melops*), goldsinny (*Ctenolabrus rupestris*), and rock cook (*Centrolabrus exoletus*), were collected in Ireland and Scotland. Most of these wrasse were screened by veterinarians, some were analysed in detail for metazoan and protozoan parasites, and the haematological characteristics of a small number determined.

Thirty-five species of parasites were recorded in this study. Differences in parasites observed between wrasse species may have reflected host specificity, diet and local distribution. The greater prevalence of ectoparasites on ballan (*Labrus bergylta*) and cuckoo (*L. mixtus*) may reflect the lack of cleaning behaviour in adult ballan and male cuckoo. There were differences in lymphocyctes and granulocytes, but not blood serum lysozyme and antiprotease concentrations, nor erythrocyte, leucocyte, and thrombocyte counts, nor haematocrits, between corkwing, goldsinny, and rock cook.

No parasites, bacteria or viruses of known pathogenic concern to salmonids were recorded from wild wrasse. Four wrasse had an atypical strain of *Aeromonas salmonicida* which was not pathogenic to salmon. The parasites and diseases of wrasse are still poorly known and further studies are needed on their taxonomy and life-history. Fish farms should only use healthy wrasse and continue to promote the study of parasites and diseases of wrasse and other wild fish associated with salmon cages.

17.1 Introduction

Until their use as cleaner-fish for the control of sea lice on Atlantic salmon (*Salmo salar* L.) farms (Costello & Bjordal, 1990), wrasse were of no commercial value in northern Europe. This change in their importance has resulted in a marked increase in research into the biology of the five common species in northern Europe, namely; ballan (*Labrus bergylta* Ascanius); cuckoo (*Labrus mixtus* L.); corkwing [*Crenilabrus melops* (L.)]; goldsinny [*Ctenolabrus rupestris* (L.)], and rock cook [*Centrolabrus exoletus* (L.)]. Information on their parasites was limited to incidental records of: copepods (Quignard, 1968; Boxshall, 1974; Dipper, 1976; Kabata, 1979; Radujkovic & Raibaut, 1989); isopods (Wilson, 1958; Potts, 1973; Trilles *et al.*, 1989); trematodes (Gibson & Bray, 1982; Radujkovic *et al.*, 1989; Radujkovic & Euzet, 1989); nematodes (Petter & Radujkovic, 1989); and protozoans (Möller & Anders, 1986; Daoudi *et al.*, 1989; Lubat *et al.*, 1989). Many of these records were from the Mediterranean (Daoudi *et al.*, 1989; Lubat *et al.*, 1989; Radujkovic *et al.*, 1989; Radujkovic & Euzet, 1989; Radujkovic & Raibaut, 1989; Trilles *et al.*, 1989), and there was no published information on bacteria or viruses associated with these wrasse.

In 'cleaner-fish technology' wild wrasse are stocked into salmon cages at ratios of between 50 and 150 salmon to one wrasse. The wrasse swim amongst the salmon and pluck lice from their skin. The intensive nature of modern fish farming facilitates the spread of pathogenic organisms and Atlantic salmon farms have suffered significantly from diseases and parasites (Costello, 1993, on sea lice). This experience led to concern that wild wrasse may harbour organisms which could infect the salmon, and resulted in additional records of copepods, trematodes, protozoans and bacteria infecting the five species of wrasse (Costello, 1991; Treasurer & Cox, 1991; Bron & Treasurer, 1992; Frerichs *et al.*, 1992; Menezes, 1992; Donnelly & Reynolds, 1994; Treasurer & Laidler, 1994). Most records related to goldsinny, corkwing and rock cook, as these were the species used as cleaner-fish. In conjunction with our studies on cleaner-fish technology in Ireland and Scotland, parasitological, microbiological and veterinary screening of these species was conducted. In addition, the opportunity was taken to examine the haematology of the species. These results are reported in this paper, and combined with a review of other records to assess the threat of wrasse parasites and diseases to farmed Atlantic salmon.

17.2   Materials and methods

In total, 822 wrasse were collected from the North Water of Mulroy Bay (55°15'N 7°30'W) in County Donegal, north-west Ireland; from the Ballinakill (53°35'N 10°5'W) and Lettercallow (53°12'N 9°20'W) Bays in County Galway, west of Ireland; from Ardgroom Bay (51°46'N 9°10'W) in County Kerry, south-west Ireland; and from the Sound of Mull (56°30'N 5°45 W) in western Scotland (Table 17.1). The low number of rock cook examined reflected their rarity in the fishery, and the high number of corkwing reflected the effort of researchers working in Mulroy Bay, where this species was unusually abundant. Wrasse from Mulroy Bay were screened in October 1989 (45 corkwing), June 1990 (150 corkwing), September 1990 (20 corkwing), and June 1991 (100 corkwing and goldsinny). The

Table 17.1 The number of each species of wrasse examined from each locality. Samples in which only parasites were examined, and those screened by fish veterinarians and microbiologically, are indicated.

	Corkwing	Goldsinny	Rock cook
Parasites only			
Lettercallow Bay	20	20	0
Sound of Mull	30	30	30
Veterinary screening			
Lettercallow Bay	43	115	0
Ballinakill Bay	21	37	51
Ardgroom Bay	3	104	3
Mulroy Bay	300	15	0
Total	417	321	84

June 1991 samples consisted of 73 male and 12 female corkwing (total length 7–17 cm) and 10 male and five female goldsinny (total length 8–16 cm). Detailed examination of parasites was conducted on samples from Lettercallow Bay and the Sound of Mull, and of protozoans from the latter area. Analysis for viruses was limited to samples from Mulroy Bay.

17.2.1   *Parasitology*

Live wrasse were selected for disease screening before being placed into cages with salmon. Initial screening of wrasse from Mulroy Bay was conducted at the on-site laboratory of Fanad Fisheries Ltd, with further examination at the Institute of Aquaculture, University of Stirling when required. Wet gill-mounts and skin scrapes were examined for parasites. Samples of skin, muscle, gills, heart, intestine, kidney and stomach were preserved in phosphate buffered 4% formaldehyde, and examined histologically at the University of Stirling.

Wrasse used in the west of Ireland were screened by veterinarians at the National Diagnostics Centre (NDC), University College, Galway, and their metazoan parasites identified by S. Deady. These wrasse were captured in the morning, immediately transported to the laboratory, and kept alive until analysis (within 48 hrs of capture).

The metazoan and protozoan parasites of wrasse from Scottish waters were identified by A. Pike and colleagues (Harvey, *unpubl. data*; Edwards, 1992; Thom, 1992); Dr I. Dykova, Institute of Parasitology, Czech Academy of Sciences, assisted with the identification of protozoans. Wrasse were captured from the Sound of Mull, collected from holding cages in Lochaline in west Scotland, and transported live to aquaria in Aberdeen. They were examined fresh, within three weeks of arrival at the laboratory, to facilitate the finding and identification of protozoans. As fish can lose intestinal parasites during starvation and holding in captivity, abundance counts of intestinal parasites from these wrasse may be underestimated. Samples of 10 goldsinny, 10 corkwing, and 10 rock cook were collected in October 1991, June 1992 and July–August 1992.

17.2.2   *Bacteriology and virology*

At Fanad, bacterial screening consisted of making smears from kidney tissue onto four media: (a) tryptone soya agar (TSA); (b) marine agar; (c) cytophaga agar; and (d) selective kidney disease medium. Media (c) and (d) are selective for myxobacteria and bacterial kidney disease respectively. Smears on media (a) to (c) were incubated at 20°C for 4 days, and on (d) at 15° C for 10 weeks. The resulting colonies were identified following Amos (1985). For virological examin-ation, kidney tissues were aseptically transported in pools of five at <5°C to the University of Stirling. Tissues were innoculated onto RTG–2 and BF–2 cells, incubated at 15°C, and examined according to Amos (1985).

   At the NDC, wrasse were killed by a blow to the head, the caudal peduncle severed, and a blood sample taken immediately in a haematocrit tube. The tube was spun at 10000 rpm for 10 min and the packed cell volume (PCV) measured on a haematocrit reader. Blood smears were also taken and stained with either pinacyanol chloride or Giemsa. A portion of gill, and a skin scrape from the dorsal fin, were placed with a drop of seawater on a glass slide, and examined microscopically for organisms. Following dissection, portions of kidney were aseptically plated onto Brain Heart Infusion Agar (BHIA), and onto BHIA with 1% NaCl. Both of these samples were then incubated for seven days at 22°C, and observed for an additional seven days. For histological analysis, portions of gill, spleen, heart, liver, kidney, pancreas and muscle tissues were fixed in buffered 4% formalin for 48 hrs. They were then routinely processed, imbedded in paraffin wax, 5 μm sections cut, and stained with haematoxylin and eosin. At the NDC, a sample of 20 goldsinny and 5 corkwing from Lettercallow Bay were subjected to a stress test used to stimulate the appearance of latent furunculosis, a bacterial disease caused by *Aeromonas salmonicida salmonicida*.

17.2.3   *Haematology*

As a first step in characterizing the immune system of wrasse, the first haemato-logical studies on north European wrasse were conducted during this project. As already described, haematocrit was routinely determined as part of veterinary screening by the NDC. In addition, blood serum was analysed and blood cells identified and enumerated at Aberdeen (Passmore, 1992).

17.3 Results

17.3.1 *Parasitology*

In this study, 35 species of parasites were recorded from corkwing, goldsinny and rock cook (Table 17.2), including 14 species of Protozoa. *Trichodinid* protozoans were common on the gills of all species from each locality. Nematodes and digenean trematodes were usually present, but monogenean trematodes were not frequent and no isopods were recorded. *Leposphilus labrei* was common in all the corkwing populations studied, but was not recorded in any other species of wrasse. Its frequency of occurrence increased with fish size and age. Uniquely amongst the parasites recorded, there was never more than one *L. labrei* per host despite its high prevalence (Table 17.3). A quantitative study of wrasse from Lettercallow Bay found some parasites to be frequent and sometimes abundant (Table 17.3); notably *Hatschekia cluthae*, cysts of *Galactosomum lacteum*, and *Lecithochirium rufoviride* on corkwing; and *Hysterothylacium aduncum* and cysts of *Cryptocotyle lingua* on goldsinny.

The locations of parasite species on corkwing, goldsinny and rock cook were similar (Table 17.4). However, these species had proportionally fewer external parasite species than given in the literature for ballan and cuckoo. The absence of protozoan parasites, and low numbers of parasites recorded from the body cavity and organs of the latter two species, probably reflects a lack of investigation. More parasites have been recorded from corkwing than from the other four species.

In addition to the detailed examination of wrasse for parasites, veterinary screening of wrasse from Mulroy Bay, Lettercallow Bay, Ballinakill Bay, and Ardgroom Bay commonly recorded *Trichodina* sp. on the gills. Monogean trematodes, *Gyrodactylus* sp., were present on the gills of some wrasse from these bays and were associated with focal hyperplasia of the epithelium. All wrasse examined appeared to be healthy, with the exception of two goldsinny damaged during capture and handling.

17.3.2 *Bacteriology and virology*

In 1991, 4500 corkwing and 320 goldsinny were stocked into the salmon cages in Mulroy Bay. According to Ossiander and Wedemeyer (1973), for a population of 5000 fish a sample of 288 fish is necessary to detect a disease with an incidence of 1.0% respectively (with 95% confidence), and a minimum of 20 fish is required to detect an incidence of 10%. Thus the screening carried out for wrasse in Mulroy Bay may be considered sufficient to detect diseases with an incidence of <1% in corkwing but only 10% for goldsinny.

All screenings of wrasse from Mulroy Bay were negative for the viral pathogens tested. The first and second screenings were negative for bacterial pathogens. In

Table 17.2 Records of parasites from the five species of wrasse occurring in northern Europe. Sources: Wilson, 1958; Quignard, 1968; Boxshall, 1974; Dipper, 1976; Potts, 1973; Kabata, 1979; Möller & Anders, 1986; Daoudi *et al.*, 1989; Lubat *et al.*, 1989; Petter & Radujkovic, 1989; Radujkovic & Euzet, 1989; Radujkovic & Raibaut, 1989; Trilles *et al.*, 1989; Bron & Treasurer, 1992; Bjordal, *unpubl. data*. (1) = Sound of Mull, 1991 and 1992; (2) = Ardgroom Bay, Co. Kerry, Ireland, 1991; (3) = Ballinakill Bay, Co. Galway, Ireland, 1991; (4) = Lettercallow Bay, Co. Galway, 1991 and 1992; x = existing record in literature; (x) = published record but only from wrasse inside salmon cages; + = new host record. Immature Allocreadidae are probably either *Gaerskajatrema perezi* or *Macricaria alacris*.

Parasite	Location on host	Ballan	Cuckoo	Corkwing	Goldsinny	Rock cook
ACANTHOCEPHALA						
Echinorhynchus gadi Zoega	intestine	–	–	+4	–	–
Polymorphus sp. (van Cleeve)	around gut	–	–	+4	–	–
Cystacanth (unidentified)	visceral cavity	–	–	+1	+1	–
CRUSTACEA: COPEPODA						
Caligus centrodonti Baird	skin	x	–	+1	x	×1
Caligus elongatus Nordmann	skin	(x)	–	–	(x)	–
Caligus labracis Scott	inside operculum	x	–	–	–	–
Colobomatus bergylta Hesse	embedded in skin	x	–	–	–	–
Hatschekia cluthae (Scott)	gills	x	–	+1,4	×1,4	–
Hatschekia labracis (van Beneden)	gills	x	x	–	–	+1
Hatschekia pygmaea Scott & Scott	gills	–	–	x	x	–
Hatschekia sp. Poche	gills	–	–	×4	×4	×2,3
Holobomolochus confusus (Stock)	nasal cavity	x	–	–	–	–
Leposphilus labrei Hesse	embedded in lateral line	–	–	×1,2,3,4	–	–
	embedded in cephalic canals	–	–	–	–	x
Lernaeocera branchialis (L.)	gills	x	–	–	–	–
CRUSTACEA: ISOPODA						
Anilocra frontalis Edwards	skin	x	–	x	–	–
Gnathia maxillaris (Montagu)	skin	x	x	–	–	–
Nerocila bivittata (Risso)	fins	–	–	x	–	–

Species	Site				
NEMATODA					
Cucullanus micropapillatus Törnquist	intestine	–	x	–	–
Contracaecum sp. Railliet & Henry	intestine, liver	–	+3,4	+2,4	+3
Hysterothylacium aduncum (Rudolphi)	liver and intestine	–	+4	+4	–
Hysterothylacium sp. Ward & Magath	visceral cavity	–	+1	+1	+1
	intestine	–	–	–	+1
	gall bladder	–	+1	–	–
Raphidascaris sp. Railliet & Henry	visceral cavity	–	+1	+1	+1
TREMATODA: DIGENEA					
Cryptocotyle lingua (Creplin)	cyst under skin	–	×1,4	×1,4	+1
Dactylogyrus sp. Diesing	gills	–	x	–	–
Gaevskajatrema perezi (Mathias)	intestine	–	+4	–	–
Galactosomum lacteum (Jagersk.)	kidney and gills	–	+4	+4	–
Galactosomum sp. Looss	kidney, peritoneum	–	+2	+2	–
Helicometra fasciata Rudolphi	intestine	×1	×1	–	–
Lepidauchen stenostoma Nicoll	intestine	x	–	–	–
Macvicaria alacris Gibson & Bray	intestine	x	×1,4	×1,4	×1
Podocotyle sp. Dujardin	intestine	–	+2	+2	–
Lecithochirium rufoviride (Rudolphi)	intestine	–	×4	×4	+1
Proctoeces sp. Odhner	intestine, lumen	–	×1	×1	+1
	gonad	–	–	–	+1
Prosorhynchus aculeatum Odhner	gills	–	+4	+4	–
Immature Allocreadidae (Rudolphi)	intestine (adults)	–	+4	–	–
Zoogonus rubellus (Olsson)					
TREMATODA: MONOGENEA					
Gyrodactylus sp. Nordmann	gills	x	–	–	+1
Microcotyle donavini Van Beneden & Hesse	gills	–	–	–	–

Table 17.2 *Continued*

Parasite	Location on host	Ballan	Cuckoo	Corkwing	Goldsinny	Rock cook
PROTOZOA: APICOMPLEXA						
Eimeria banyulensis Lom & Dykova	intestine	–	–	–	x	–
Eimeria sp. Schneider	intestinal mucosa	–	–	+1	+1	–
Goussia sp. Labbe	intestine lumen	–	–	+4	+4	–
PROTOZOA: CILIOPHORA						
Trichodina sp. Ehrenberg	skin and gills	–	–	+1,2,3,4	×1,2,3,4	+1,2,3
Trichodina ovonucleata Raabe	gills	–	–	+1	–	–
Trichodina rectuncinata Raabe	gills	–	–	+1	–	–
Trichodina labrorum Chatton	gills	–	–	x	–	–
Trichodina sp.1	gills	–	–	–	+1	+1
Trichodina sp.2	gills	–	–	+1	–	–
PROTOZOA: MICROSPORA						
Microsporida indetermined	muscle and hepato-pancreas	–	–	×3,4	×4	–
Icthyosporidium giganteum (Thélohan)	connective tissue	–	–	×	–	–
PROTOZOA: MYXOSPOREA						
Sphaerospora divergens Thélohan	intestine	–	–	+1	–	–
Davisia sp. (Davis)	urinary bladder	–	–	–	+1	–
Ortholinea divergens (Thélohan)	urinary bladder	–	–	–	–	+1
PROTOZOA: ZOOMASTIGOPHORA						
Bodonidae (similar to *Cryptobia* sp.)	gills	–	–	+2	+2,4	+3
Ichthyobodo sp. (Henneguy)	gills	–	–	+1	–	+1
Number records to species level		12	4	20	8	7
Additional taxa (e.g. genera different from those above)		0	0	8	5	5
Total number of taxa		12	4	28	13	12

Table 17.3 Counts of metazoan parasites on 13 male and seven female corkwing (age birth to five years); and 17 male, one female, and two unsexed goldsinny (age three to six years), all from Lettercallow Bay, (see Table 2 for sites of infection). Data are presented as percentage of fish infested, total number of parasites on all fish , and range of parasites per fish.

Parasite	Corkwing			Goldsinny		
	Abundance		% infested	Abundance		% infested
	Total	Range		Total	Range	
Echinorhychus gadi	3	0–3	5	0	0	0
Polymorphus sp.	2	0–1	10	0	0	0
Hatschekia cluthae	419	0–74	75	6	0–3	20
Hysterothylacium aduncum	10	0–5	20	693	2–152	100
Leposphilus labrei	10	0–1	50	0	0	0
Cryptocotyle lingua	58	0–17	45	**	0–>200	95
Gaevskjatrema perezi	44	0–20	25	0	0	0
Galactostomum lacteum cysts	*	0–>200	95	10	0–3	15
Macvicaria alacris	14	0–10	15	3	0–1	15
Immature Allocreadidae	31	0–12	25	5	0–2	20
Prosorhynchus aculeatus	11	0–5	30	0	0	0
Lecithochirium rufoviride	276	0–58	95	62	0–25	55

* most fish had 10 to 50 cysts each.
** most fish had 20 to 100 cysts each.

Table 17.4 Relative occurrence (as a percentage) of the parasites listed in Table 17.2 in different parts of the wrasse host.

Parasite location	% Occurrence				
	Ballan	Cuckoo	Corkwing	Goldsinny	Rock cook
Skin, fins, and inside operculum	33	25	10	4	10
Intestine	17	50	33	36	25
Body cavity and body organs	0	0	24	32	25
Gills	33	25	29	25	30
Inside skin, nasal cavity and cephalic canals	17	0	5	4	10

July 1990, corkwing in the salmon cages at Moross became infected by a *Vibrio* sp. bacterium a few days after the salmon became infected. After administering antibiotics in the salmon diet the salmon recovered, and shortly afterwards the wrasse also recovered. In September 1990, nine of 20 wild captured corkwing had significant levels of *Vibrio* sp., as determined by a confluent bacterial monoculture. The widespread bacterium *Aeromonas hydrophila* was isolated from some corkwing kidney smears but not considered of pathological importance.

In the first week of June 1991, *Vibrio* sp. was isolated from 13%, and an atypical *Aeromonas salmonicida* from 4% of a sample of 85 corkwing and 15 goldsinny wrasse. Unfortunately the sample labels did not identify the wrasse species or sex. All four isolates of *Aeromonas* were chemically indistinguishable (following tests in Amos, 1985); cultures on TSA were unusually slow growing, did not show characteristic brown pigment until seven to eight days of growth, and were smaller than normal *A. salmonicida* isolates. At the University of Stirling, the atypical *Aeromonas* was found to be sensitive to the antibiotics oxytetracycline, oxolinic acid, potentiated sulphonamide, furazolidone, and nitro-furantoin, but resistant to amoxycillin. Analysis at the NDC similarly found sensitivity to oxytetracycline, oxolinic acid, and furazolidone, but found resistance to cotrimoxazole (a potentiated sulphonamide) and kanamycin. Strains of this bacterium can cause the highly pathogenic disease furunculosis in salmon. In view of the presence of the atypical *Aeromonas salmonicida*, the fish farm discontinued its use of wrasse so as not to prejudice its sales of salmon eggs and smolts to other farms through any association with a disease. In subsequent tests Frerichs *et al.* (1992) found the bacterium to be non-pathogenic to salmon. In their paper the species of wrasse was mistakenly given as goldsinny, whereas the samples were taken from a mixed (i.e. species not identified) sample of 15 goldsinny and 85 corkwing.

Bacterial screening of goldsinny and corkwing from Lettercallow Bay, including a furunculosis stress test in 1991, proved negative, with the exception of some tentatively identified *Chlamydia* sp. on the gills which was not considered of pathological concern. Similarly, bacterial screening of goldsinny, corkwing, and rock cook from Ardgroom and Ballinakill Bays proved negative. In Lettercallow Bay, 1561 goldsinny and 452 corkwing were stocked in 1991 and 1992. At these population sizes, the numbers of wrasse tested would be sufficient to detect a carrier with 5% and 10% prevalence respectively (Ossiander & Wedemeyer, 1973).

17.3.3   *Haematology*

There were no significant differences in the serum lysozyme and antiprotease, and erythrocyte, leucocyte and thrombocyte counts of goldsinny, rock cook and corkwing collected from the Sound of Mull (Table 17.5). However, rock cook had significantly less lymphocytes and more type 1 granulocytes than either of the other species, and more type 2 granulocytes than corkwing. Haematocrits of wrasse from Lettercallow Bay were 39.2 ±2.1 (PCV, mean ± 95% CL) for goldsinny ($n = 56$) and 43.1 ±4.5 for corkwing ($n = 23$). Levels ranged from 22% to 60% PCV for goldsinny, and 15% to 61% for corkwing.

Table 17.5 Blood serum and cell characteristics of corkwing, goldsinny and rock cook wrasse collected in the west of Scotland. Results as mean ± SD, with sample size in parenthesis. Data from Passmore (1992).

Parameter	Goldsinny	Rock cook	Corkwing
Serum lysozyme ($\mu g\,ml^{-1}$)	3.95 ± 2.79 (5)	<0.62 ± 0 (4)	1.72 ± 0.22 (3)
Serum antiprotease (units of trypsin inhibited, $\mu g\,ml^{-1}$)	17.54 ± 5.00 (4)	17.86 ± 4.77 (3)	14.59 ± 2.11 (3)
Erythrocytes (ml^{-1} serum)	$1.17 \times 10^9 \pm 0.27$ (5)	$1.81 \times 10^9 \pm 0.60$ (5)	$3.37 \times 10^9 \pm 0.92$ (5)
Leucocytes (ml^{-1} serum)	$0.24 \times 10^6 \pm 0.24$ (5)	$1.33 \times 10^6 \pm 1.28$ (5)	$0.87 \times 10^6 \pm 0.87$ (5)
Lymphocytes %	87.3 ± 5.9 (4)	57.0 ± 10.1 (5)	74.4 ± 10.2 (5)
Granulocytes	(4)	(5)	(5)
type I %	1.5 ± 1.0	15.2 ± 7.5	8.0 ± 5.6
type II %	6.3 ± 4.7	11.4 ± 7.2	1.0 ± 1.7
Thrombocytes %	1.3 ± 1.9 (4)	6.6 ± 5.2 (5)	5.0 ± 4.6 (5)
Unidentified %	3.7 ± 3.3 (4)	9.6 ± 8.4 (5)	11.6 ± 9.2 (5)

17.4   Discussion

17.4.1   *Parasitology*

At least 42 species of parasites are now known from the five species of wrasse in northern Europe (Table 17.2). Sampling of 111 *Symphodus tinca* (L.) and 97 *Labrus merula* L. from the western Mediterranean found 18 and 17 metazoan species of parasites, including the 14 species listed in Table 17.2 (Campos & Carbonell, 1994). A relatively small number of fish have been examined from few localities, and it is likely that further studies will discover new parasites on the species. A more comprehensive literature review than presented here, particularly on studies from around the turn of the century, has found additional records of parasites on these wrasse species (Karlsbakk, *unpubl.*). However, these records do not alter the patterns discussed here.

The greater proportion of external parasites on ballan and cuckoo may reflect the lack of cleaning behaviour in ballan and male cuckoo (this volume, Chapter 13). The other species show intraspecific cleaning behaviour. Although corkwing appear to harbour more parasite species, this may reflect sampling effort rather than a true biological phenomenon. Indeed, the results of this study, and the literature review, indicate that knowledge of wrasse parasites is still rudimentary. Even for the populations examined in this study, either the number of fish (e.g. in the Sound of Mull) or the scope of investigation (e.g. for Protozoa) was too limited.

There were contrasting abundances of the parasites *H. cluthae* and *G. lacteum* on corkwing, and *H. aduncum* and *C. lingua* on goldsinny, in Lettercallow Bay (Table 17.3). *Hatschekia cluthae*, and *H. labracis* and *H. pygmaea*, appear specific to north European Labridae (Kabata, 1979), and transmission is probably direct by the free-swimming larvae. The larvae and adults of the nematode *H. aduncum* inhabit a variety of invertebrate and fish species with transmission to wrasse through predation. The digenean trematodes *G. lacteum* and *C. lingua* infest littorinid gastropods (as rediae larvae), a wide range of fish species (as metacercarial cysts), and seagulls (as adults), with transmission to fish probably by free-swimming larvae. The differences in intermediate hosts and modes of transmission of parasites will interact with the host species diets and local distribution to result in the observed patterns of infestation. Future studies would benefit by examining the host diet to identify possible intermediate hosts of parasites.

Most of the recorded parasites are likely to be specific to wrasse, or to require the wrasse either to be eaten by another fish or other animal, or to pass to an invertebrate host, to complete their life-cycle (Table 17.6). The occasional possibility of a salmon eating a wrasse is unlikely to result in a pathological problem to the salmon stock because firstly the load will be low (i.e. was not pathogenic to the wrasse) to the salmon, secondly the parasite may not survive in the salmon; and thirdly further transmission between salmon (through intraspecific

Table 17.6 Mode of transmission, intermediate and final hosts (if any), and site of infection of wrasse parasites. The potential of wrasse parasites to transfer to salmon within sea-cages, is indicated by: *unlikely or only if salmon eats wrasse; **transfer via water possible but pathogenic effect unlikely due to host specificity or normally of commensal nature; ***possible transfer and pathogenic effect; ? = uncertainty.

Species	Transmission	Intermediate hosts	Final hosts	Site of infection	Transfer potential
ACANTHOCEPHALA	predation	invertebrate	vertebrate	intestine	*
CRUSTACEA					
Copepoda	water	none	fish	gills, skin	**
Isopoda	free-swimming	none	fish	skin	*
NEMATODA	predation	invertebrate	vertebrate	intestine, viscera	*
TREMATODA					
Monogenea	water	none	fish	gills	**
Digenea	predation	invertebrate	vertebrate	intestine	*
PROTOZOA					
Apicomplexa	water?	none?	fish	intracellular	***?
Ciliophora	water?	none	fish	gills, skin	**
Microspora	water?	none?	fish	intracellular	***?
Myxosporea	water?	none	fish	urinary and gall bladders	***
Zoomastigophora	water	invertebrate	fish	gills, skin	**

predation) is improbable. Even if transmitted, nematodes (though sometimes present in high numbers), monogeneans, and ciliophora are not normally pathogenic (Rodgers, *pers. comm.*). The trichodonids, bodonids, and micro-sporidea are frequently present on, but not pathogenic to, farmed salmonids (Rodgers, *pers. comm.*).

Of more concern to the salmon are parasites with the potential for direct transmission between hosts, and which may be viable on salmon. Most of the crustacean (with the exception of *Lernaeocera branchialis*) and protozoan parasites, and the monogenean trematodes are not known to require intermediate hosts (Table 17.6). In contrast to trematodes and protozoans, crustacean parasites are often pathogenic when on cultured fish. However, most of the copepod crustaceans (such as *Hatschekia* sp., *Leposphilus* sp.) are host-specific, with the exception of *Caligus elongatus*. However, the latter species has not been recorded on wild wrasse and was only recorded on one goldsinny in salmon cages (Bron & Treasurer, 1992). The isopod adults do not feed, they live on the seabed, and their larvae swim up to attack fish. Larval densities are rarely high and wrasse and salmon would be equally likely to attack. Although the method of transmission may not be well understood for protozoans and monogeneans, and they may not be known as salmonid pathogens, it is prudent to screen wrasse to determine their presence. In this way, should an unusual parasite attack salmon, the role of wrasse as a vector could be evaluated.

17.4.2 *Bacteriology and virology*

During this study, 271 goldsinny, 367 corkwing, and 54 rock cook were given a veterinary screening, and most of these were tested for the presence of pathogenic bacteria. All screenings proved negative for viral and bacterial pathogens, with the exception of an atypical strain of *Aeromonas salmonicida* identified from four of a 100-wrasse sample (85 corkwing and 15 goldsinny) from Mulroy Bay. The *Vibrio* sp. and *Aeromonas hydrophila* bacteria detected on wrasse at Fanad regularly occured on the salmon. They were considered widespread bacteria of moderate to low pathogenicity, which were only apparent when the fish became stressed. However, in laboratory tests Gravningen *et al.* (this volume, Chapter 20) found goldsinny to be killed by injection of *Vibrio anguillarum* strains, *A. salmonicida salmonicida* (the disease furunculosis), and atypical *Aeromonas salmonicida*, within days. Wrasse feeding on dead salmon which had contracted furunculosis later died from the disease (Treasurer & Laidler, 1994), so both the injected and the consumed bacteria transferred the infection to wrasse.

Strains of *A. salmonicida* occur widely in marine fish and wild salmonids (Willumsen, 1990; Menezes, *pers. comm.*), and in salmonids effects range from non-pathogenic to highly virulent (Hussein, 1991). The strain identified from the wrasse proved to be non-pathogenic to Atlantic salmon (Frerichs *et al.*, 1992). An atypical strain of *A. salmonicida* was pathogenic to wrasse (mainly goldsinny) in salmon cages in Norway, but the strain was not detected in the salmon present (this volume, Chapter 15). There is always a probability, however remote, that wrasse may harbour a microbial pathogen transmissible and harmful to salmon. Such a probability is likely to be decreased if wrasse were cultured in a disease-free hatchery. If such disease-free cultured wrasse become available this may encourage more salmon farms to use wrasse in lice control.

17.4.3   *Haematology*

The haematological parameters studied may be affected by stress, be it due to capture or husbandry conditions. As only five fish of each species were examined in Scotland, further studies are necessary to determine the biological significance of the results.

The transmission of *A. salmonicida* from Atlantic salmon to goldsinny and rock cook in salmon cages has shown wrasse to be sensitive to this widespread salmonid pathogen (Treasurer & Cox, 1991; Treasurer & Laidler, 1994). The trials in Mulroy Bay in 1990 found wrasse to be susceptible to *Vibrio* infections although mortalities were not as high as for the salmon; the salmon were probably more stressed at the high summer temperatures. It may therefore be desirable to vaccinate wrasse as well as smolts (Treasurer & Laidler, 1994; this volume, Chapter 20). Further information on the haematology and immunology of wrasse will be needed to test the suitability of vaccines for wrasse. Additionally, should wrasse culture be undertaken, an understanding of the normal levels of haemato-logical parameters will enable the state of health of the wrasse to be determined.

17.5   **Conclusions**

A proportion of wrasse from populations of each species stocked at fish farms should always be subject to veterinary examination. No wrasse pathogen has yet been identified which is of particular concern to either the wrasse or salmon. A survey by the Scottish Office marine laboratory similarly found no evidence for a significant threat to salmon health from wild wrasse being used as cleaner-fish, though many parasites occurred on the wrasse (SOAFD, 1994). However, only healthy wrasse should be used in salmon farms, as wrasse in poor condition can develop heavy (perhaps debilitating) parasite infections (this volume, Chapter 22). It is notable that many other species of fish are closely associated with salmon cages, including mackerel, *Scomber scombrus* L., grey mullet, *Chelon labrosus* Risso, saithe, *Pollachius virens* (L.), and pollack, *P. pollachius* (L.) (Mason & Costello, *unpubl. data*). Saithe have been suspected of being transient hosts of sea lice (Bruno & Stone, 1990), and unlike the territorial wrasse, saithe travel between farm sites. In the absence of a particular pathogen problem to focus research on wrasse, it may be more appropriate to examine the parasites and diseases of all fish species regularly associated with salmon cages. Such studies should pay particular attention to: identifying the transmission mechanisms of bacteria, viruses, protozoans, and metazoans; and, to determining the pathogenicity of selected species to wild fish, wrasse, and cultured salmonids. The results of normal veterinary screening are not sufficient for such research, as it is critical that the parasites be identified to species level, thereby entailing considerable taxonomic effort.

Acknowledgements

This study was funded by contract AQ.2.502 under the Fisheries and Aquaculture Research programme of the European Union. We thank Ms Catherine McManus and the other staff at Fanad Fisheries Ltd, and the staff of other fish farms for their cooperation. Mr Douglas Lamont kindly donated wrasse from the Sound of Mull to this study. Mr Will Darwall, Ms Siobhán Lysaght, Ms Denise McCorry, Mr Richard Donnelly (all of TCD), and Ms Sarah Varian (UCG), also assisted this study. Dr Hamish Rodger and Dr Evelyn Collins (NDC, UCG) provided helpful comments in addition to screening wrasse for the project. We thank Mr Egil Karlsbakk (University of Bergen) for providing information on records of parasites of wrasse.

References

Amos, K.H. (1985) *Procedures for the Detection and Identification of Certain Fish Pathogens.* Fish Health Section, American Fisheries Society, 3rd edn, Corvallis, Oregon.

Boxshall, G.A. (1974) Infections with parasitic copepods in North Sea marine fishes. *Journal of the Marine Biological Association of the United Kingdom* **54**: 355–72.

Bron, J.E. and Treasurer, J.W. (1992) Sea lice (Caligidae) on wrasse (Labridae) from selected British wild and salmon-farm sources. *Journal of the Marine Biological Association of the United Kingdom* **72**: 645–50.

Bruno, D.W. and Stone, J. (1990) The role of saithe, *Pollachius virens* L., as a host for sea lice, *Lepeophtheirus salmonis* Krøyer and *Caligus elongatus* Nordmann. *Aquaculture* **89**: 201–7.

Campos, A. and Carbonell, E. (1994) Parasite community diversity in two Mediterranean labrid fishes *Symphodus tinca* and *Labrus merula*. *Journal of Fish Biology* **44**: 409–13.

Costello, M.J. (1991) Review of the biology of wrasse (Labridae: Pisces) in Northern Europe. *Progress in Underwater Science* **16**: 29–51.

Costello, M.J. (1993) Review of methods to control sea lice (Caligidae, Crustacea) infestations on salmon farms. In *Pathogens of Wild and Farmed Fish: Sea Lice* (Ed. by Boxshall, G.A. and Defaye, D.). Ellis Horwood Ltd, Chicester, pp. 219–52.

Costello, M.J. and Bjordal, Å. (1990) How good is this natural control of sea lice? *Fish Farmer* **13(3)**: 44–6.

Daoudi, F., Radujkovic, B.M., Marques, A. and Bouix, G. (1989) Parasites des poissons marins du Monténégro: Coccidies. *Acta Adriatica* **30**: 13–30.

Dipper, F.A. (1976) Reproductive biology of Manx Labridae. Ph.D. thesis, University of Liverpool.

Donnelly, R.E. and Reynolds, J.D. (1994) Occurrence and distribution of the parasitic copepod *Leposphilus labrei* on corkwing wrasse (*Crenilabrus melops*) from Mulroy Bay, Ireland. *Journal of Parasitology* **80**: 331–2.

Edwards, J. (1992) The parasitology of saithe (*Pollachus virens*) and wrasse (Labridae) and the possible transmission into salmonid aquaculture. Unpublished MSc thesis, University of Aberdeen.

Frerichs, G.N., Millar, S.D. and McManus, C. (1992) Atypical *Aeromonas salmonicida* isolated from healthy wrasse (*Ctenolabrus ruprestris*). *Bulletin of the European Association of Fish Pathologists* **12**: 48–9.

Gibson, D.I. and Bray, R.A. (1982) A study and reorganisation of *Plagioporus* Stafford, 1904 (Digena: Opecoelidae) and related genera, with special reference to forms from European Atlantic waters. *Journal of Natural History* **16**: 529–59.

Hussein, M.S. (1991) Identifying virulence of *Aeromonas salmonicida* strains to Atlantic salmon, *Salmo salar*, in vivo. Unpublished MSc thesis, University of Stirling.

Kabata, Z. (1979) *Parasitic Copepoda of British fishes*. The Ray Society, London.

Lubat, V., Radujkovic, B.M., Marques, A. and Bouix, G. (1989) Parasites des poissons marins du Monténégro: Myxosporidies. *Acta Adriatica* **30**: 31–50.

Menezes, J. (1992) Hazards from pathogens carried by wild fish particularly wrasse used as a lice cleaner. *Bulletin of the European Association of Fish Pathologists* **12**: 194–5.

Möller, H. and Anders, K. (1986) *Diseases and Parasites of Marine Fishes*. Möller, Kiel.

Ossiander, F.J. and Wedemeyer, G. (1973) Computer programme for sample sizes required to determine disease incidence in fish populations. *Journal of the Fisheries Research Board of Canada* **30**: 1383–4.

Passmore, E. (1992) A preliminary investigation of the wrasse (Labridae) immune system. Unpublished MSc thesis, University of Aberdeen.

Petter, A.J. and Radujkovic, B.M. (1989) Parasites des poissons marins du Monténégro: Nématodes. *Acta Adriatica* **30**: 195–236.

Potts, G.W. (1973) Cleaning symbiosis among British fish with special reference to *Crenilabrus melops* (Labridae). *Journal of the Marine Biological Association of the United Kingdom* **53**: 1–10.

Quignard, J.-P. (1968) Rapport entre la présence d'une 'gibbosité frontale' chez les Labridae (Poissons, Téléostéens) et le parasite *Leposphilus labrei* Hesse, 1866 (Copépode Philichthyidae). *Annales Parasitologie* (Paris) **43**: 51–7.

Radujkovic, B.M. and Euzet, L. (1989) Parasites des poissons marins du Monténégro: Monogenes. *Acta Adriatica* **30**: 51–135.

Radujkovic, B.M. and Raibaut, A. (1989) Parasites des poissons marins du Monténégro: Copepodes. *Acta Adriatica* **30**: 237–78.

Radujkovic, B.M., Orecchia, P. and Paggi, L. (1989) Parasites des poissons marins du Monténégro: Digenes. *Acta Adriatica* **30**: 137–87.

Scottish Office Agriculture and Fisheries Department (1994) Marine Laboratory Aberdeen Annual review 1992–93. SOAFD, Aberdeen.

Thom, K.G. (1992) Parasites of wrasse (*Centrolabrus exoletus*, *Crenilabrus melops* and *Ctenolabrus rupestris*) in relation to their possible transmission to Atlantic salmon (*Salmo salar*). Unpublished BSc thesis, University of Aberdeen.

Treasurer, J. and Cox, D. (1991) The occurrence of *Aeromonas salmonicida* in wrasse (Labridae) and implications for Atlantic salmon farming. *Bulletin of the European Association of Fish Pathologists* **11**: 208–10.

Treasurer, J. and Laidler, L.A. (1994) *Aeromonas salmonicida* infection in wrasse (Labridae), used as cleaner fish, on an Atlantic salmon, *Salmo salar* L., farm. *Journal of Fish Diseases* **17**: 155–61.

Trilles, J.-P., Radujkovic, B.M. and Romestand, B. (1989) Parasites des poissons marins du Monténégro: Isopodes. *Acta Adriatica* **30**: 279–306.

Willumsen, B. (1990) *Aeromonas salmonicida* subsp. *salmonicida* isolated from Atlantic cod and coalfish. *Bulletin of the European Association of Fish Pathologists* **10**: 62–3.

Wilson, D.P. (1958) Notes from the Plymouth aquarium. III. *Journal of the Marine Biological Association of the United Kingdom* **37**: 299–307.

Chapter 18

Health status of goldsinny wrasse, including a detailed examination of the parasite community at Flødevigen, southern Norway

E. KARLSBAKK, K. HODNELAND and A. NYLUND *Department of Fisheries and Marine Biology, University of Bergen, Bergen High Technology Centre, N–5020 Bergen, Norway*

A screening of disease agents of goldsinny wrasse, *Ctenolabrus rupestris*, from Flødevigen, southern Norway, was undertaken to reveal their health status, and as potential carriers of wrasse and salmon diseases. No viral or bacterial agents were discovered, and the parasite species revealed were either monogenetic (one-host life-cycle) wrasse specialists, or larval forms of generalists. The parasite community of goldsinny was depauperate, both in species numbers and abundance. Compared with Mediterranean labrid species, the metazoan community was less diverse, particularly because of a lower abundance of ectoparasitic crustaceans, and a depauperate intestinal helminth community.

18.1   Introduction

Several species of wrasse (Labridae) are used as cleaner-fish to remove caligid copepods on farmed Atlantic salmon, *Salmo salar* L., in Ireland, Scotland and Norway (Costello & Bjordal, 1990). A new commercial fishery has developed in these countries, to supply salmon farms with live wrasse. In Norway, mainly goldsinny [*Ctenolabrus rupestris* (L.)] are used, but also corkwing [*Crenilabrus melops* (L.)] and rock cook [*Centrolabrus exoletus* (L.)]. These species have their northernmost limits of distribution in Norway (Quignard & Pras, 1986). Losses of wrasse in salmon pens due to disease and winter mortality, and the low numbers found in northern Norway, means that transportation of wrasse is necessary, particularly from the south to the north of the country (Kvenseth, 1993).

Movements of live fish always involve the risk of introduction of new species or strains of pathogens (Stewart, 1991). Transport of live wrasse may involve intro-duction of wrasse pathogens to naïve local wrasse populations, salmon diseases to new farms and areas, and diseases of wild fish to new areas. The main problem for the individual farmer is the survival expectancy of wrasse in salmon pens, and their potential as carriers of salmon diseases. The present study was undertaken to examine the health status of a shipment of goldsinny wrasse from Flødevigen near Arendal, southern Norway, to Bodø in Nordland, northern Norway. The goldsinny were examined for bacteria, IPN virus (IPNV) and eukaryotic micro- and macroparasites. The structure of the parasite community was examined in more detail to allow comparison with parasite communities in other Labridae.

18.2   Materials and methods

Approximately 20 000 wrasse (99.5% goldsinny) were supplied in July 1994 by local fishermen, and held in tanks for two weeks at the research station, Institute of Marine Research, in Flødevigen, southern Norway (58°25′N, 08°44′E). Fish were held in flowing seawater at 13°C, and were not fed. A subsample of 320 goldsinny was studied, and the remaining wrasse were transported in tanks to Bodø, where they were distributed to several salmon farms. A total of 59 goldsinny were each killed by a blow to the head and examined fresh for ectosymbiotic microparasites, with scrapings of skin, fins and the first left gill. An additional 260 goldsinny were killed by prolonged anaesthesia (MS–222), and transported in individual plastic bags on ice to the laboratory. Of these, 60 were immediately deep-frozen for later parasitological study, while the remaining 200 were screened for bacteria and IPNV using sterile techniques. All fish were measured to total length and weighed (Table 18.1).

Kidney samples for IPNV screening were homogenized and diluted in EBSS (Flow). After centrifugation (4500 g, 15 min) the supernatant was passed through a 0.22 μm filter and inoculated on CHSE–214 cell monolayer (1:100). The cells were grown at 20°C and were examined daily for cytopathic effect (CPE).

Innoculum for bacteria were taken from the middle region of the kidney, and inoculated on blood agar plates with or without 2% sodium chloride. Two series were grown at 15 and 22°C, and examined after 48 and 96 hours. Bacterial colonies from positive plates were re-innoculated and sent to the State Veterinary Laboratory in Bergen for identification.

The parasitological study included examination of skin, fins and gills for ectoparasites and trematode metacercariae, and musculature and viscera for helminths. Microscopical examination was performed on the gall bladder, urinary bladder and posterior kidney of all fish. Microscopical examination of the central nervous system (CNS) (medulla oblongata), and intestinal mucosa were performed

Table 18.1 Parameters of goldsinny in subsamples from Flødevigen, southern Norway (*n* = sample size).

Sample	Bacteria and virus	Parasites
n	200	58
Length (mm) mean ± SD (range)	117.4 ± 10.2 (90–147)	116.8 ± 12.1 (95–145)
Weight (g) mean ± SD (range)	22.6 ± 6.6 (9–43)	21.7 ± 7.7 (12–42)
Age mode (range)	–	4 (4–6)
% females	37.0	38.6

on 30 fish. The terminology used is as defined by Margolis *et al.* (1982) and Esch *et al.* (1990). Component species (prevalence $\geq$10%) were used in accordance with Bush *et al.* (1990). When parametric statistics were used, the data were transformed by log (x) or log (x + 1) using base 2 or 10 logarithms, which normalized the data. Brillouin's diversity index for fully censused communities was calculated, as described in Pielou (1975) using natural logarithms.

18.3   Results

18.3.1   *Virology and bacteriology*

All IPNV tests were negative. Three bacterial isolates were re-inoculated and studied, but did not prove to conform to known ichthyopathogenic species, and hence were not characterized further.

18.3.2   *Parasite fauna*

A total of 17 parasite species were identified. Of these, four were protozoan species, and 13 were metazoans (Table 18.2).

Protozoa

All protozoans were ectosymbiotic ciliates or flagellates. The trichodinid *Paratrichodina* sp. dominated the protozoan gill infections with frequent massive aggregations (>100 ciliates per primary lamella). These ciliates covered the secondary lamellae of the gills, but did not occur on the skin. A *Trichodina* species (type 1) also infected the gills. An infrequent *Trichodina* (type 2) occurred at very low intensity on the skin. The flagellate *Cryptobia* sp. adhered to the secondary lamellae in moderate intensities (<10 per primary lamella), its presence or absence was not related to the trichodinid intensities. Microscopical examination of the gall bladder, urinary bladder, kidney, intestinal mucosa and medulla oblongata did not reveal spore-producing protozoans. Myxosporean plasmodia of the *Ceratomyxa* type were encountered in the gall bladder in one instance.

Trematoda

Metacercarial cysts of *Cryptocotyle lingua* occurred particularly on the fins, with the pectoral fins and the caudal fin harbouring the majority of the infrapopulations. *Lecithochirium* metacercariae were encapsulated on the serosal surface of all visceral organs and in the mesenteries (81.2%). Some metacercariae were also found in the musculature (18.8%), particularly the areas surrounding the visceral cavity (i.e. the belly flaps). The metacercariae were often clustered in masses and located on the anterior intestine or in the space between the intestine

Table 18.2 Parasites of goldsinny from Flødevigen, southern Norway (protozoa, $n = 59$; metazoa, $n = 58$).

Major taxon/Species	Site	Prevalence (%)	Abundance ± SD[1]	Intensity range (mean ± SD)
Sarcomastigophora				
Cryptobia sp. Leidy	Gills	11.9	–	–
Ciliophora				
Paratrichodina sp. Lom	Gills	98.3	–	–
Trichodina sp. I	Gills	40.7	–	–
Trichodina sp. II	Skin	6.8	–	–
Trematoda				
Cryptocotyle lingua (Creplin) metacercariae	Skin, fins, gills	56.1	1.7 ± 2.9	1–16 (3.1 ± 3.4)
Lecithochirium sp. Luehe metacercariae	Liver, intestine, musculature	100.0	77.4 ± 97.8	1–484 (77.4 ± 97.8)
Derogenes varicus (Mueller)	Anterior intestine	1.7	(0.02 ± 0.13)	1 (1)
Lecithaster gibbosus (Rudolphi)	Intestine	1.7	(0.02 ± 0.13)	1 (1)
Cestoda				
Grillotia erinaceus Guiart plerocercoid	On intestine	1.7	(0.02 ± 0.13)	1 (1)
Nematoda				
Cosmocephalus obvelatus (Creplin) III larvae	Mesenteries	36.2	0.9 ± 2.0	1–13 (2.4 ± 2.8)
Paracuaria adunca (Creplin) III larvae	Mesenteries	24.1	0.5 ± 1.2	1–8 (1.9 ± 1.9)
Hysterothylacium aduncum (Rudolphi) III larvae	Mesenteries and on intestine	100.0	17.8 ± 15.6	1–104 (17.8 ± 15.6)
Hysterothylacium aduncum IV/V	Intestinal lumen	1.7	(0.03 ± 0.26)	2 (2)
Contracaecum septentrionalis Railliet & Henry III larvae	Mesenteries	3.4	(0.03 ± 0.18)	1 (1)
Acanthocephala				
Echinorhynchus gadi (Zoega)	Intestine	5.2	(0.05 ± 0.22)	1 (1)
Corynosoma semerme Rudolphi cystacanth	Viscera	5.2	(0.07 ± 0.32)	1–2 (1.3 ± 0.6)
Copepoda				
Hatschekia cluthae (Scolt)	Gills	1.7	(0.02 ± 0.13)	1 (1)
Caligus centrodonti Baird	Skin and fins	1.7	(0.02 ± 0.13)	1 (1)

[1] Abundance given in parentheses for intensities ≤2.

and the liver, or around the anal part of the intestine. There was also a tendency for metacercariae to be found in the capsules of the larval nematode *Hysterothylacium aduncum*. Degenerate metacercariae of all sizes were common on the viscera. Of adult gastrointestinal trematodes, a single specimen of both *Derogenes varicus* and *Lecithaster gibbosus* were found. These were both oviferous, although not fully developed.

Cestoda

One single, fully developed plerocercoid of the trypanorhynch *Grillotia erinaceus* was found encapsulated in the posterior visceral cavity.

Nematoda

Small third stage larvae (<3 mm total length) of the acuariid nematodes *Cosmocephalus obvelatus* and *Paracuaria adunca* occurred coiled in small capsules in the intestinal wall and the mesenteries. Third stage larvae of the anisakid *Hysterothylacium aduncum* were encapsulated on the intestine and in the mesenteries of the digestive tract and adjoining organs. Degenerate specimens were common.

Acanthocephala

Cystacanths of *Corynosoma* occurred encapsulated on the digestive tract. A total of three specimens of *Echinorhynchus gadi* was recovered from the intestine; all were small males.

Copepoda

Specimens of two species (one of each) were recovered. An adult oviferous female *Hatschekia cluthae* was removed from the first gill arch, and a fourth chalimus of *Caligus centrodonti* from the caudal fin.

18.3.3   *Relationship between component species and host parameters*

Infections with metazoan component species were tested against host parameters, including length, age, sex, and condition (Table 18.3); and against each other for association and correlation. Parasitic infection with *Lecithochirium* sp. metacercariae decreased significantly with increasing host length ($P < 0.05$) and female goldsinny had significantly higher intensities (124.3, 1 SD = 131.6) than males (48.6, 1 SD = 55.4) ($P < 0.05$). The length composition of the sexes in the sample did not differ significantly ($P > 0.05$). *Lecithochirium* sp. intensity was negatively correlated with host condition ($P < 0.05$), while *P. adunca* was positively

Table 18.3 Relationship between helminth component species and host parameters of goldsinny from Flødevigen, southern Norway (*P < 0.05; **P < 0.01).

Species	LENGTH Spearman[1] r_S	AGE Spearman[1] r_S	SEX Student's *t*-test[2] t	CONDITION[3] Linear regression r (F)
Crytocotyle lingua	−0.084	0.032	0.524	−0.008 (0.003)
Lecithochirium sp.	−0.279*	0.037	2.346*	−0.291 (5.200)*
Hysterothylacium aduncum III	−0.160	0.060	1.402	−0.215 (2.711)
Paracuaria adunca III	0.285*	0.402**	0.116	0.215 (2.726)
Cosmocephalus obvelatus III	0.234	0.161	0.081	0.190 (2.107)

[1] Spearman's rank correlation coefficient, (r_S).
[2] Student's *t*-test on $\log_{10}(x + 1)$ transformed abundance data.
[3] Fulton's condition factor.

correlated although not significantly. A significant positive correlation was observed between the intensities of *H. aduncum* III and *Lecithochirium* sp. metacercariae (Spearman's rank correlation, $r_s = 0.439$, $P < 0.001$). The acuariid larvae, *P. adunca* and *C. obvelatus* were associated ($\chi^2 = 9.91$, DF = 1, $P < 0.01$), although not correlated.

18.3.4   *Parasite community*

Prevalence values reveal three core species (prevalence >70%) in the parasite community. These were the ciliate ectosymbiont *Paratrichodina*, and the two larval helminths *Lecithochirium* sp. and *Hysterothylacium aduncum*. Eleven satellite species were recognized (prevalence <30%), and these included two copepods *Hatschekia cluthae* and *Caligus centrodonti*, the only host (or labrid) specialist metazoans. The remaining species had intermediate prevalences (secondary species). All the metazoan component species were generalists. The status of the trichodinids is unknown, although the gill trichodinids are probably restricted to labrids (Karlsbakk, *pers. obs.*). *Cryptobia* sp. is known from a range of unrelated littoral fishes in Norway (Karlsbakk *et al.*, 1994).

The metazoan parasite community was dominated by larval helminths, representing eight of 13 species, and 86–100% of all metazoan individuals. Larval helminths represented 100% of all metazoan individuals in 87.9% of the component communities. Of the larval helminth species, five were allogenic and three autogenic species. Allogenic species use aquatic vertebrates as intermediate hosts and mature in birds or mammals, while autogenic species complete their entire life-cycle in the aquatic environment (Esch *et al.*, 1990). The autogenic species constituted 49–100% of all larval helminth individuals; they represented 100% of all helminth larvae in 31% of the fish, and 100% of all metazoans in 25.8% of the fish. All the allogenic parasites identified are species which mature in fish-eating birds.

Table 18.4 Number of parasite species, individuals and values of Brillouin's diversity index (H) for infracommunities in goldsinny from Flødevigen, southern Norway.

Measure	Metazoan community		Intestinal helminth community	
	mean $\pm$ 1 SD	range	mean $\pm$ 1 SD	range
Number of species	3.22 $\pm$ 1.01	2–6	0.10 $\pm$ 0.36	0–2
Number of individuals	97.91 $\pm$ 106.00	10–527	0.12 $\pm$ 0.46	0–3
Brillouin's H	0.579 $\pm$ 0.266	0.076–1.372	0.073–0.164	0–0.366
Proportion of samples with 0 or 1 species	0.000		0.983	

The diversity parameters are summarized in Table 18.4. There was a positive relationship between host length and diversity (Spearman's rank correlation, $r_s = 0.308$, $P < 0.05$), while there were no significant correlations between Brillouin's H and age, condition or sex of goldsinny ($P < 0.05$ in each case). Of the parasites, the metazoan diversity (H) only correlated with the *Lecithochirium* sp. intensity (Spearman's rank correlation, $r_s = -0.511$, $P < 0.001$). More diverse infracommunities contained few of these larval trematodes. The number of metazoan parasite species increased significantly with the age of the fish host (Spearman's rank correlation, $r_s = 0.284$, $P < 0.05$), but not with length.

The intestinal helminth fauna comprised a total of four species, all of which (treating intestinal *H. aduncum* separately) were generalists and satellite species. Prevalence of intestinal helminths was 12%, with a total of nine helminth individuals.

18.4   Discussion

18.4.1   *Health status and implications*

Since commencement of use in salmon culture, mortality of wrasse in cages has led to bacterial screening of both stocked and free-living populations. Mortalities caused by *Aeromonas salmonicida* ssp. *salmonicida* have been observed in Scotland when wrasse were stocked with salmon suffering from furunculosis (Collins *et al.*, 1991; Treasurer & Cox, 1991; Treasurer & Laidler, 1994). Atypical *A. salmonicida* has been isolated from goldsinny in both natural populations (Frerichs *et al.*, 1992), and from goldsinny stocked with salmon as cleaner-fish (Viken & Ektvedt, 1993; Kvenseth, *pers. comm.*). No fish-pathogenic bacteria were isolated from goldsinny in Flødevigen. Treasurer & Laidler (1994) advised the use of stress testing of the wrasse to reveal carriers. This was not attempted in the present study, but it is assumed that the wrasse experienced some degree of stress during capture, transport and tank holding.

Viral agents have not been detected in goldsinny or other wrasse species.

However, negative virus tests, as carried out by Treasurer (1991), Treasurer & Cox (1991) and in this study, do not mean lack of wrasse viruses, but are rather a reflection of the lack of investigations. Gibson & Sommerville (this volume, Chapter 19) demonstrated experimental infection with IPNV specific serotype in goldsinny wrasse.

The Flødevigen area has no salmon farming, and therefore is suitable for detection of natural wrasse disease agents. Of particular interest is the presence of atypical *A. salmonicida* in wrasse, and the possibility that this agent may be enzootic in Norwegian wrasse. Further screening of wrasse in non-farming areas may help elucidate this problem.

Potential parasitic salmon diseases carried by wrasse should be monogenetic species (direct life-cycles). These agents may be ectoparasites such as copepods, monogeneans, ciliates, flagellates and amoebae, and endoparasites such as intestinal flagellates. No such agents were identified in the present study as the copepods from goldsinny are specific to labrids. The trichodinids of goldsinny appear to be labrid specialists (Karlsbakk, *pers. obs.*), and are certainly distinct from the *Trichodina* species occurring on the gills of salmon in the sea (*T. californica* Davis, according to Shulman & Shulman-Albova, 1953). *Cryptobia* sp. is, although unspecific with regard to host, considered as a harmless commensal (Karlsbakk *et al.*, 1994). Wrasse may transmit intestinal helminths such as the nematode *Hysterothylacium aduncum* if eaten by salmon, but this is probably not significant as salmon is a common host of this nematode in the wild (Shulman & Shulman-Albova, 1953; Bristow & Berland, 1991). The results from the present study suggest a good health status for goldsinny from Flødevigen, and that these fish are not carriers of salmon parasites. There are reservations, however, regarding wrasse as potential latent carriers of viral and bacterial diseases. Goldsinny may also be host to amoebae and intestinal flagellates, although these were not discovered in the present study.

18.4.2   *The metazoan parasites*

The component metazoan parasite species of goldsinny were all larval helminths, suggesting that the host is an important transmitter rather than final host. The metacercariae of *Cryptocotyle lingua* present in the skin appear as black spots and infect most teleost fish (Möller & Anders, 1986). The cercariae emerge from rediae in *Littorina* spp. [mainly *L. littorea* (L.)] in spring and summer, and actively penetrate the host epidermis, where they encyst (Stunkard, 1930). Final hosts are mainly birds (Möller & Anders, 1986).

The acuariid nematode larvae, *Paracuaria adunca* and *Cosmocephalus obvelatus* also mature in birds (gulls). The life-cycles of the two species are similar, with amphipods and mysids as first intermediate hosts (Anderson & Wong, 1982; Wong & Anderson, 1982; Marcogliese, 1992). Fishes become infected when eating amphipods or mysids, and act as parathenic hosts (Anderson & Wong,

1982; Wong & Anderson, 1982). Associated infections by these related nematodes may be accomplished through their use of common intermediate hosts. However, the marine intermediate hosts in the northeast Atlantic are unknown.

The nematode *Hysterothylacium aduncum* matures in most teleost fishes (Möller & Anders, 1986). Various smaller crustaceans, particularly copepods, amphipods, isopods and mysids, are first intermediate hosts. Invertebrate predators and fish act as second intermediate or parathenic hosts (Køie, 1993). Whether fish such as goldsinny become infected with *H. aduncum* at the fourth (IV) and adult (V) stage in the intestine, or as third (III) stage larvae encapsulated on the viscera, depends on the size of the larvae in the intermediate host. Larvae less than *c.* 3 mm re-encapsulate, while larger larvae may moult and establish luminal infections (*cf.* Køie, 1993). The scarcity of stage IV and V *H. aduncum* in the goldsinny relative to stage III, may indicate that infections are acquired mainly from small copepods. As the goldsinny is predominantly a benthic feeder, these are probably harpacticoids, but the diet may also contain planktonic copepods (Hilldén, 1978).

The *Lecithochirium* sp. metacercariae were similar to *L. rufoviride* (Rudolphi) from the common eel, *Anguilla anguilla* L. *Lecithochirium* metacercariae from wrasse have been reported under several names, and reports of *Synaptobothrium caudiporum* (Rudolphi) metacercariae from wrasse is probably also *Lecithochirium* (Gibson & Bray, 1986). Køie (1984) found goldsinny from the Faroes infected with *Lecithochirium* metacercariae, thought to be *L. rufoviride* from the common eel. First intermediate hosts are gastropods of genus *Gibbula*, while crustaceans are second intermediate hosts. As with *H. aduncum*, the number of obligatory hosts in the life-cycle depends on the size attained by the parasite in the second intermediate host. A four-host cycle results when copepods act as intermediate hosts, while a three-host cycle is achieved through amphipods and isopods (Køie, 1990). A negative relationship of *Lecithochirium* abundance with increasing host-length, suggests that invertebrates, acting as intermediate hosts, have a decreasing importance in the diet of goldsinny with increasing length, and this may apply to copepods (Sayer *et al.*, 1995). Degenerate metacercariae were common, and thus the observed tendency may be accomplished through mortality in the infra-populations. The intensity of metacercariae was also correlated with *H. aduncum* intensity, which may indicate that these parasites benefit from their mutual presence. The occurrence of *Lecithochirium* metacercariae in the *H. aduncum* capsules may be such an example. However, the correlation of these parasites may also be accomplished through common intermediate hosts.

18.4.3   *Parasite community structure*

There have been few studies on the parasite community diversity in marine fish. Kennedy *et al.* (1986) compared the intestinal helminth community diversity in several freshwater fishes and birds, and showed that birds supported diverse communities, while fish were comparatively depauperate in species and numbers of individuals. They predicted that, due to a higher invertebrate diversity in the sea (potential intermediate hosts), helminth communities of marine fishes should be more diverse than those of freshwater fishes. Recent studies by Kennedy and Williams (1989), Holmes (1990), Thoney (1993) and Campos & Carbonell (1994) support this view. The latter authors studied the metazoan parasite community of two labrid species, *Symphodus tinca* (L.) and *Labrus merula* L. *S. tinca* had 18 parasite species and *L. merula* 17, with 11 common to both. Goldsinny at Flødevigen had fewer species (13), but the difference is clearer when the number of component species are compared. Compared to five species in goldsinny, *S. tinca* and *L. merula* had ten and nine species respectively. The metazoan parasite communities were also more diverse in the Mediterranean species, with a mean of 4.15 and 4.17 parasite species, and Brillouin indexes of 0.904 and 1.022 respectively (Campos & Carbonell, 1994).

Two important differences between the parasite fauna of the labrids studied by Campos & Carbonell (1994) and the goldsinny in this study were evident. First, there is a higher number of intestinal helminth and ectoparasitic crustacean species in the Mediterranean labrids; and, second, their abundances are higher. By restricting analysis in the present study to intestinal helminths, the first explanation for the depauperacy of the goldsinny metazoan parasite community in Flødevigen is confirmed. A diversity analysis is scarcely meaningful, as 91.4% of the fish had no helminths; also the Brillouin index, the mean number of helminth species and individuals were very low, even compared to freshwater fishes (Kennedy *et al.*, 1986). The apparent scarcity of intestinal helminths in goldsinny was also noted by Nicoll (1907).

Kennedy *et al.* (1986) considered five factors essential for the production of diverse intestinal helminth communities:

(i)   complexity of host alimentary canal (producing different environments or sites) and physiology (ecto-/endothermy);
(ii)   host vagility (movement with time; mobility);
(iii) broad host diet;
(iv)   selective feeding on particular intermediate hosts;
(v)   exposure to monogenetic species which enter by penetration. Criterion (v) is not relevant in fish (Kennedy *et al.*, 1986).

Goldsinny possess a simple alimentary tract with no clear distinction between a stomach and an intestine, and also has a lack of caeca (Sayer *et al.*, 1995). Criterion (i) would support a depauperate community in goldsinny, with fewer niches for colonization. Host vagility (criterion (ii)) is also restricted in goldsinny wrasse, which are territorial (Hilldén, 1981; this volume, Chapter 5). This may influence the diet (criterion (iii)), and, with a patchy distribution of invertebrate intermediate hosts, contribute to selective feeding (criterion (iv)) and helminth

over-dispersion. The diet is fairly broad, but dominated by epibenthic invertebrates (Hilldén, 1978; Sayer *et al.*, 1995). Most common were crustaceans
(amphipods and barnacles), polychaetes, gastropods, bivalves, hydrozoans
(Scotland only) and bryozoans (Sweden only). A broad diet should contribute to
a diverse helminth community. However, apart from the two cited studies (in
Sweden and Scotland), the diet of this species is not well known, and local conditions in Flødevigen may well cause the particular scarcity of intestinal helminths
in goldsinny at that locality. Some invertebrate groups identified to be important
by Hilldén (1978) and Sayer *et al.* (1995), such as hydrozoa and the epiphytic
bryozoa, are also unknown as intermediate hosts for helminths. Considering the
total metazoan parasite community, the amphipods are obligatory or potential
intermediate hosts for all the component species identified in this study, except *C.
lingua*, and for some satellite species, such as the acanthocephalans. Preliminary
results from examination of other wrasse species–corkwing, rock cook, ballan
(*Labrus bergylta* Ascanius) and cuckoo (*L. mixtus* L.)–from western Norway,
suggest that these support diverse intestinal helminth communities, with trematodes of the family Opecoelidae being particularly abundant (Karlsbakk, *unpubl.
data*). These observations therefore tend to undermine the importance of criterion
(i), as these labrid species have a simple alimentary tract (Quignard, 1966).

Acknowledgements

This study was financed by P.G. Kvenseth, of A/S MOWI.

References

Anderson, R.C. and Wong, P.L. (1982) The transmission and development of *Paracuaria adunca*
    (Creplin, 1846) (Nematoda: Acuarioidea) of gulls (Laridae). *Canadian Journal of Zoology* **60**:
    3092–104.
Bristow, G.A. and Berland, B. (1991) A report on some metazoan parasites of wild marine salmon
    (*Salmo salar* L.) from the west coast of Norway with comments on their interactions with farmed
    salmon. *Aquaculture* **98**: 311–18.
Bush, A.O., Aho, J.M. and Kennedy, C.R. (1990) Ecological versus phylogenetic determinants of
    helminth parasite community richness. *Evolutionary Ecology* **4**: 1–20.
Campos, A. and Carbonell, E. (1994) Parasite community structure in two Mediterranean labrid fishes
    Symphodus tinca and *Labrus merula*. *Journal of Fish Biology* **44**: 409–13.
Collins, R.O., Ferguson, D.A. and Bonniwell, M.A. (1991) Furunculosis in wrasse. *Veterinary
    Record*. **128**: 43.
Costello, M.J. and Bjordal, Å. (1990) How good is this natural control on sea lice? *Fish Farmer* **13(3)**:
    44–6.
Esch, G.W., Shostak, A.W., Marcogliese, D.J. and Goater, T.M. (1990) Patterns and processes in
    helminth communities: an overview. In *Parasite Communities: Patterns and Processes* (Ed. by
    Esch, G.W., Bush, A.O. and Aho, J.M.). Chapman and Hall, London, pp. 1–19.
Frerichs, G.N., Millar, S.D. and McManus, C. (1992) Atypical *Aeromonas salmonicida* isolated from
    healthy wrasse (*Ctenolabrus rupestris*). *Bulletin of the European Association of Fish Pathologists*
    12: 48–9.
Gibson, D.I. and Bray, R.A. (1986) The Hemiuridae (Digenea) of fishes from the north-east Atlantic.
    Bulletin of the British Museum (Natural History) (Zoology) **51**: 1–125.

Hilldén, N.-O. (1978) On the feeding of the goldsinny, *Ctenolabrus rupestris* L. (Pisces, Labridae). *Ophelia* **17**: 195–8.

Hilldén, N.-O. (1981) Territoriality and reproductive behaviour in the goldsinny *Ctenolabrus rupestris* L. *Behavioural Processes* **6**: 207–21.

Holmes, J.C. (1990) Helminth communities in marine fishes. In *Parasite Communities: Patterns and Processes* (Ed. by Esch, G.W., Bush, A.O. and Aho, J.M.). Chapman and Hall, London, pp. 101–30.

Karlsbakk, E., Nilsen, F. and Hodneland, K. (1994) Ecto- and endosymbiotic flagellates of the genus *Cryptobia* infecting Norwegian marine fishes. *International Symposium on Aquatic Animal Health*, Seattle, September 4–8, 1994, P–44.

Kennedy, C.R., Bush, A.O. and Aho, J.M. (1986) Patterns in helminth communities: why are birds and fish so different? *Parasitology* **93**: 205–15.

Kennedy, C.R. and Williams, H.H. (1989) Helminth parasite community diversity in a marine fish *Raja batis* L. *Journal of Fish Biology* **34**: 971–2.

Kvenseth, P.G. (1993) Wrasse culture–soon a reality? *Norsk Fiskeoppdrett* **18**: 28–9 (in Norwegian).

Køie, M. (1984) Digenetic trematodes from *Gadus morhua* L. (Osteichthyes, Gadidae) from Danish waters, with special reference to their life-histories. *Ophelia* **23**: 195–222.

Køie, M. (1990) Redescription of the cercaria of *Lecithochirium rufoviride* (Rudolphi, 1819) Lühe, 1901 (Digenea, Hemiuridae) (= *Cercaria vaullegeardi* Pelseneer, 1906). *Ophelia* **31**: 85–95.

Køie, M. (1993) Aspects of the life cycle and morphology of *Hysterothylacium aduncum* (Rudolphi, 1802) (Nematoda, Ascaridoidea, Anisakidae). *Canadian Journal of Zoology* **71**: 1289–96.

Marcogliese, D.J. (1992) *Neomysis americana* (Crustacea: Mysidacea) as an intermediate host for sealworm, *Pseudoterranova decipiens* (Nematoda: Ascaridoidea), and spirurid nematodes (Acuarioidea). *Canadian Journal of Fisheries and Aquatic Sciences* **49**: 513–15.

Margolis, L., Esch, G.W., Holmes, J.C., Kuris, A.M. and Schad, G.A. (1982) The use of ecological terms in parasitology (report of an Ad Hoc Committee of the American Society of Parasitologists). *Journal of Parasitology* **68**: 131–3.

Möller, H. and Anders, K. (1986) *Diseases and Parasites of Marine Fishes*. Verlag Heino Möller, Kiel.

Nicoll, W. (1907) A contribution towards a knowledge of the Entozoa of British marine fishes, Part 1. *Annals and Magazine of Natural History* (Series 7) **19**: 66–94.

Pielou, E.C. (1975) *Ecological Diversity*. Wiley–Interscience, New York.

Quignard, J.-P. (1966) Recherches sur les Labridae (Poissons, Teleostéens, Perciformes) des côtes européennes: systématique et biologie. *Naturalia Monspeliensia (Zoologie)* **5**: 7–248.

Quignard, J.-P. and Pras, A. (1986) Labridae. In *Fishes of the North-eastern Atlantic and the Mediterranean* Vol. II (Ed. by Whitehead, P.J.P., Bauchot, M.-L., Hureau, J.-C., Nielson, J. and Tortonese, E.). UNESCO, Paris, pp. 919–42.

Sayer, M.D.J., Gibson, R.N. and Atkinson, R.J.A. (1995) Growth, diet and condition of goldsinny on the west coast of Scotland. *Journal of Fish Biology* **46**: 317–40.

Shulman, S.S. and Shulman-Albova, R.E. (1953) *Parasites of Fishes of the White Sea*. Izdatelstvo Akademia Nauk SSSR, Moscow (in Russian).

Stewart, J.E. (1991) Introductions as factors in disease of fish and aquatic invertebrates. *Canadian Journal of Fisheries and Aquatic Sciences* (Supplement 1) **48**: 110–17.

Stunkard, H.W. (1930) The life history of *Cryptocotyle lingua* (Creplin), with notes on the physiology of the metacercariae. *Journal of Morphology* **50**: 143–90.

Thoney, D.A. (1993) Community ecology of the parasites of adult spot, *Leiostomus xanthurus*, and Atlantic croaker, *Micropogonias undulatus* (Sciaenidae) in the Cape Hatteras region. *Journal of Fish Biology* **43**: 781–804.

Treasurer, J. (1991) Limitations in the use of wrasse. *Fish Farmer* **14(5)**: 12–13.

Treasurer, J.W. and Cox, D. (1991) The occurrence of *Aeromonas salmonicida* in wrasse (Labridae) and implications for Atlantic salmon farming. *Bulletin of the European Association of Fish Pathologists* **11**: 208–10.

Treasurer, J.W. and Laidler, L.A. (1994) *Aeromonas salmonicida* infection in wrasse (Labridae), used as cleaner fish, on an Atlantic salmon, *Salmo salar* L., farm. *Journal of Fish Diseases* **17**: 155–61.

Viken, A. and Ektvedt, R. (1993) Increased mortality of wrasse in salmon farms–a casuistic. *Akvavet* **3**: 33–5 (in Norwegian).

Wong, P.L. and Anderson, R.C. (1982) The transmission and development of *Cosmocephalus obvelatus* (Nematoda: Acuarioidea) of gulls (Laridae). *Canadian Journal of Zoology* **60**: 1426–40.

Chapter 19
The potential for viral problems related to the use of wrasse in the farming of Atlantic salmon

D.R. GIBSON and C. SOMMERVILLE *Institute of Aquaculture, University of Stirling, Stirling FK9 4LA, Scotland, UK*

The use of wrasse (Labridae) to control the numbers of the parasitic copepods *Caligus elongatus* and *Lepeoptheirus salmonis* (sea lice) is now widespread in the farming of Atlantic salmon *(Salmo salar)*. Despite large numbers of wrasse being caught from the wild in Scotland for use in this way there have been no reported cases of viral diseases being imported with them. In view of the important nature of viral diseases in the farming of salmon a review of the literature pertaining to viral diseases in wrasse was undertaken and experiments were carried out involving the two most important viral diseases affecting Scottish salmon production. These diseases are infectious pancreatic necrosis (IPN) and pancreas disease (PD). Cultured goldsinny (*Ctenolabrus rupestris*), aged two years, were used in both studies. The goldsinny used were found to be susceptible to IPN at dose levels comparable to those infective to salmon. In the study, goldsinny were shown to possess a greater recovery ability from the disease when compared with salmon. Fish from the same group were found to be unaffected by PD at dose levels infective to salmon. The susceptibility of goldsinny to IPNV suggests that importation of infected fish into a salmon farm site could result in the imposition of a restriction order under the Diseases of Fish Act 1937, 1983.

19.1 Introduction

The use of wrasse (Labridae) for the biological control of the parasitic copepods *Caligus elongatus* Nordmann and *Lepeoptheirus salmonis* Krøyer (sea lice) in the farming of *Salmo salar* L. (Atlantic salmon) has brought many advantages, the majority of which are environmental. It has however, introduced several new problems. These include the possible introduction of diseases with the importation of wild caught wrasse or the moving of wrasse between farm sites. For example an atypical strain of *Aeromonas salmonicida* has been isolated from wild caught wrasse in Ireland (Frerichs *et al.*, 1992) and a second isolation of atypical *A. salmonicida* is reported by Gravningen *et al.* (this volume, Chapter 20). Although this strain was not pathogenic to salmon, it emphasises the possibility of the

240

introduction of previously non-endemic pathogens to cultured salmon. The introduction of new viral pathogens would create a particularly difficult problem to overcome. Though control of some viral conditions of higher vertebrates has been made possible by the development of vaccines this approach appears to be of little use for fish viruses and despite considerable research into the subject this method is not used commercially, at present, for fish viruses. This situation may well change in the future. Wolf (1988) and Roberts (1989) have provided good reviews of vaccine research in relation to fish viral conditions. There are currently no reliable chemical treatments available commercially for viral conditions of fish and thus the main method for controlling viral diseases is by the use of a variety of management strategies aimed at preventing the introduction and outbreak of the diseases, and ameliorating their effect by reducing stress inducing factors. The improvement of general fish health is also very important in the control of disease outbreaks.

19.2   Literature review

A review of the literature reveals that no viruses have been reported from any of the wrasse species used in the farming of salmon. However, it seems unlikely that wrasse do not carry viruses, or are not susceptible to viruses; the absence of information is likely to be due to the lack of interest in wrasse from a disease viewpoint, prior to their use as biological control measures. Until now, wrasse have been of no economic value and there has been very little interest in them as experimental subjects. As far as we are aware the routine screening of wild caught wrasse for viruses does not take place prior to stocking in pens. We are aware of only two cases where wild caught wrasse were screened prior to being stocked with salmon. One of these screenings was carried out by Marine Harvest Ltd for IPN virus (IPNV) on fish caught from several different sites around the western coast of Britain (Treasurer & Cox, 1991). These sites ranged from Holyhead in Wales to the Firth of Clyde on the west of Scotland and to Gairloch and Torridon on the north west of Scotland. The species tested were goldsinny [*Ctenolabrus rupestris* (L.)], rock cook [*Centrolabrus exoletus* (L.)], corkwing [*Crenilabrus melops* (L.)] and cuckoo (*Labrus mixtus* L.). No virus was isolated from any of the fish (Treasurer, *pers. comm.*). The same results were obtained by the authors when a sample of 40 goldsinny from Loch Sunart, western Scotland, were screened for IPNV and no virus was isolated from any of the fish. However, there were an estimated 100 000 wrasse caught for use in the Scottish salmon farming industry in 1991 alone (Treasurer & Henderson, 1992) and there have been no reported cases of disease outbreaks in wrasse or salmon due to viruses imported with the wrasse. It is therefore unlikely that wild wrasse carry viruses that affect salmon or cause disease in wrasse in the farm environment.

It is difficult to monitor the progress of wrasse once stocked in cages. The tendency of wrasse to inhabit the lower reaches of cages makes them difficult to observe without the use of SCUBA-diving techniques or underwater cameras. The cost of both these methods is prohibitive for routine use. Mortalities can only be assessed when the nets are raised to facilitate the removal of salmon mortalities. By this time it is usually too late to determine the cause of death for wrasse. Reliable histological samples require to be taken within minutes of the post mortem, due to tissue degradation, and it is only a matter of hours in salt water before the corpse is unsuitable for bacteriological or virological samples. Small numbers of virus related deaths could be overlooked due to the difficulty in diagnosing the exact cause of death. For example, in a recent investigation into the causes of mortalities of wrasse in experimental salmon pens, the majority of mortalities (55%) were attributed to *Aeromonas salmonicida* (Treasurer & Laidler, 1994). However, 15% of the mortalities could not be attributed to a single cause.

In addition to the risk of importing viruses with wrasse, the susceptibility of wrasse to known salmon viral pathogens must be considered as wrasse represent a considerable investment in time and money when they are stocked in salmon pens.

19.3   Infectious pancreatic necrosis virus (IPNV)

IPNV is a known salmonid pathogen which has been well studied and is notifiable in Scotland under the Diseases of Fish Act 1937, 1983. Consequently, the import of IPNV-infected wrasse to a salmon site would result in the site being subject to restriction orders as outlined in the Act.

The effect of IPNV in salmon is well documented (Hill, 1978; Swanson & Gillespie, 1979; Smail *et al.*, 1986; Knott & Munro, 1986; Munro & Smail, 1992). The clinical signs of the disease in salmonids include darkening of the infected fish, distended abdomens and, most characteristic, a spiralling swimming motion in severely affected fish. There may also be exophthalmia. Internally there is often enteritis caused by sloughed intestinal epithelial cells. There may be signs of petechial haemorrhages over the anterior viscera accompanied by enlargement of the spleen and liver, which are often pale. The histopathological signs of the disease consist of severe focal necrosis of the exocrine pancreas with nuclear pyknosis, karyorrhexis and occasional intracytoplasmic inclusions. IPNV has also been isolated from a wide range of other fish species from both fresh and saltwater and from several piscivorous birds as well as some marine invertebrates (Sonstegarde & McDermott, 1972; Hill, 1976; Munro *et al.*, 1976; Underwood *et al.*, 1977; Ahne, 1978; Adair & Ferguson, 1981; Hudson *et al.*, 1981; Diamant *et al.*, 1988; Lo *et al.*, 1990; McAllister & Owens, 1992). The ability of the virus to survive in different salinities has also been demonstrated (Toranzo *et al.*, 1983). In view of the wide range of hosts reported for this virus, the susceptibility of wrasse to IPNV was studied experimentally by Gibson *et al.* (1995).

Cultured goldsinny, of approximately two years of age, were exposed, via bathing, to two different levels of virus. A low dose, 10^5 plaque-forming units per millilitre (pfu/ml) for one hour, and a high dose, 10^6 pfu/ml for five hours, were used. Control groups were also used. The low-dose experiment was carried out between 12 and 13°C and the high dose between the temperatures 8 and 9°C. Subsamples of the experimental population were taken at one week intervals post-infection with faeces also being sampled in the high-dose experiment.

For the low dose experiment, low tissue titres were seen soon after infection, quickly dropping to an undetectable level. In the high dose experiment, titres were higher after infection but also quickly declined. Virus was also isolated from the faeces of infected fish in the high dose experiment. Pathological changes were only seen in the experimental fish from the high dose experiment and the pancreatic necrosis seen was typical of that seen in infected salmonids. The authors stated that the results of these experiments suggested that the susceptibility of goldsinny to IPNV was similar to, if not greater than, that seen in salmonids. But the results also suggested that goldsinny possess a greater ability to recover from IPNV infection than salmonids. The presence of virus in the faeces of infected fish from the high-dose experiment indicated the ability of *C. rupestris* wrasse infected by bathing to shed virus and therefore to act as a continual source of IPNV infection in salmon farms affected by the virus.

19.4 Pancreas disease (PD)

Another disease of salmon culture systems which is considered to have a viral aetiology but is not notifiable is pancreas disease (PD) (McVicar, 1987, 1990). This disease is a major cause of losses in salmon farming. The source of infection is not known and the pathological agent has not yet been isolated. The possibility that wild fish species, such as wrasse, are the source of infection cannot be ruled out. It is possible to induce the disease in salmon by injecting naïve fish with homogenized, filtered kidney from infected fish (Raynard & Houghton, 1993). The disease is characterized in salmon by total necrosis of the exocrine pancreas accompanied by loss of feeding response and emaciation. Clinical diagnosis is often hindered by the presence of secondary pathological problems, not exclusively or consistently associated with pancreas disease. It is most likely that it is these secondary problems which cause mortalities. Viral-like inclusion bodies have been reported from affected salmon (McVicar, 1990), but it is not known if these bodies are of viral origin, and therefore their relevance to the disease cannot be confirmed (Raynard & Houghton, 1993). The severity of the outbreak is important to the rate of recovery. This recovery may only take two weeks, or may take as long as three months. The most seriously affected fish lose weight quickly, become emaciated and 'eel-like', and variable numbers of these fish never recover. No reliable indicator of when and where outbreaks occur is available (McVicar, 1986).

By 1987 the disease had been reported from 20% of Scottish salmon farm sites and had also been reported from Ireland (McVicar, 1987). The losses accountable to PD can be put down to the reduced growth rate during infection and the loss of affected fish through runting. The loss of feeding response also makes the treatment of secondary infections with oral antibiotics difficult leading to increased mortalities. There are no effective control measures although the cessation of feeding as soon as a loss of feeding response is detected and a general reduction in environmental stressors may lessen the impact of the disease outbreaks (McVicar, 1987). As was the case with IPNV, the effect of PD on wrasse must be considered.

An experiment was carried out by the authors in which cultured, disease-free goldsinny were injected with infective kidney homogenate obtained from PD-infected salmon. The method used followed the protocol laid out by Raynard and Houghton (1993). Two infective doses were used, a medium dose and a high dose, which were equivalent to those used in salmon by Raynard and Houghton (1993). Appropriate replicates and control groups were also used. In each group, 50 goldsinny were cohabited with 25 naïve salmon for the five-week duration of the experiment. Sampling of goldsinny and salmon was carried out at weekly intervals post-injection. Samples of all visceral organs were taken and processed for histology for both salmon and goldsinny at the time of sampling. During the five-week duration of the experiment, the experimental fish displayed no abnormal pathological signs. Similarly, no abnormal pathology was recorded from the cohabiting salmon. All the fish, both goldsinny and salmon, in the experiment continued to feed normally for the five-week duration of the experiment. Homogenized samples of kidney and spleen from the injected goldsinny were then injected into naïve salmon parr. None of the salmon used displayed any of the typical histopathological signs indicative of pancreas disease in salmon.

From these results, it would appear that goldsinny are refractive to the infective agent of pancreas disease and do not pass the disease on to cohabited salmon. The failure to transfer pancreas disease to naïve salmon with the injection of goldsinny kidney homogenates also indicates that the infective agent of pancreas disease is deactivated, or becomes inactive, in goldsinny injected with infective kidney homogenates from *S. salar*. Subsequent experiments with the original homogenate injected into the goldsinny showed that the transmissible agent was indeed present (Houghton *pers. comm.*).

19.5   Conclusion

It would appear that viral infections do not represent a major concern in the use of goldsinny as biological control measures for sea lice in salmon farming. The only possible restriction in their use is that they should not be moved between sites or used for the cleaning of subsequent year classes due to the possibility of goldsinny acting as carriers of IPNV. A continued screening programme should be maintained for goldsinny entering salmon pens to minimize the risk of importation of disease agents. The diseases of the wrasse species being used are still relatively unknown, and therefore precautions should be taken to prevent the introduction of new pathogens into salmon farming culture systems.

Acknowledgements

We thank Dr. A. McVicar for his invaluable assistance in this work and NERC for funding the project as part of a CASE Studentship award. We are also indebted to Dr. D. Smail, Dr. G. Houghton and Dr. R. Raynard of SOAFD, Aberdeen for their help with the experimental work.

References

Ahne, W. (1978) Isolation and characterisation of Infectious Pancreatic Necrosis Virus from Pike (*Esox lucius*). *Archives of Virology* **58**: 65–9.

Adair, B.M. and Ferguson, H.W. (1981) Isolation of Infectious Pancreatic Necrosis (IPN) Virus from non-salmonid fish. *Journal of Fish Diseases* **4**: 69–76.

Diamant, A., Smail, D.A., McFarlane, L. and Thomson, A.M. (1988) An Infectious Pancreatic Virus isolated from Common Dab *Limanda limanda*. Previously affected with X-cell disease, a disease apparently unrelated to the presence of the virus. *Diseases of Aquatic Organisms* **4**: 223–7.

Frerichs, G.N., Millar, S.D. and McManus, C. (1992) Atypical *Aeromonas salmonicida* isolated from healthy wrasse (*Ctenolabrus rupestris*). *Bulletin of the European Association of Fish Pathologists* **12**: 48–9.

Gibson, D.R., Smail, D.A. and Sommerville, C. (1995) Infectious Pancreatic Necrosis Virus: Experimental Infection of *Ctenolabrus rupestris* L. (Labridae). *Journal of Fish Diseases* (in press).

Hill, B. (1976) IPN disease. *Fisheries Management*. **7**: 90–91.

Hill, B.J. (1978) Infectious Pancreatic Necrosis Virus and its virulence. In *Microbial Diseases of Fish* (Ed. by Roberts, R.J.). Special Publication of the Society of General Microbiology, pp. 91–114.

Hudson, E.B., Bucke, D. and Forrest, A. (1981) Isolation of Infectious Pancreatic Necrosis Virus from eels, *Anguilla anguilla* L., in the United Kingdom. *Journal of Fish Diseases* **4**: 429–31.

Knott, R.M. and Munro, A.L.S. (1986) The persistence of Infectious Pancreatic Necrosis Virus in Atlantic Salmon. *Veterinary Immunology And Immunopathology* **12**: 359–64.

Lo, C-F., Lin, M-S., Liu, S-E., Wang, C-H. and Kou, G-H. (1990) Viral interference in TO-2 cells infected with IPN virus isolated from clam, *Meretrix lusoria*. *Fish Pathology* **25**: 133–40.

McAllister, P.E. and Owens, W.J. (1992) Recovery of infectious pancreatic necrosis virus from the faeces of wild piscivorous birds. *Aquaculture* **106**: 227–32.

McVicar, A.H. (1986) A spreading threat to salmon. *Fish Farmer* **9(4)**: 18–19.

McVicar, A.H. (1987) Pancreas disease of farmed Atlantic salmon, *Salmo salar*, in Scotland: Epidemiology and early pathology. *Aquaculture* **67**: 71–8.

McVicar, A.H. (1990) Infection as a primary cause of pancreas disease in farmed Atlantic salmon. *Bulletin of the European Association of Fish Pathologists* **10**: 84–7.

Munro, A.L.S., Liversidge, J. and Elson, K.G.R. (1976) The distribution of infectious pancreatic necrosis virus in wild fish in Loch Awe. *Proceeding of the Royal Society of Edinburgh (B)* **75**: 223–32.

Munro, A.L.S. and Smail, D.A. (1992) IPN Virus in farmed Atlantic salmon in Scotland. *International Council for the Exploration of the Sea, Mariculture Committee* ICES CM 1992/F: 11.

Raynard, R.S. and Houghton, G. (1993) Development towards an experimental protocol for the transmission of Pancreas Disease of Atlantic salmon *Salmo salar*. *Diseases of Aquatic Organisms* **15**: 123–8.

Roberts, R.J. (Ed.) (1989) *Fish Pathology*, 2nd edn. Bailiére Tindall, London.

Smail, D.A., Greirson, R.J. and Munro, A.L.S. (1986) Infectious Pancreatic Necrosis (IPN) Virus in Atlantic salmon: Virulence studies and sub-clinical effects with respect to growth, smolting performance and condition. *International Council for the Exploration of the Sea, Mariculture Committee* ICES CM 1986/F: 8.

Sonstegard, R.A. and MacDermott, L.A. (1972) Epidemiological model for passive transfer of IPNV by homeotherms. *Nature* **237**: 104–5.

Swanson, R.N. and Gillespie, J.H. (1979) Pathogenesis of Infectious Pancreatic Necrosis in Atlantic salmon. *Journal of the Fisheries Research Board of Canada.* **36**: 587–91.

Toranzo, A.E., Barja, J.L., Lemos, M.L. and Hetrick, F.M. (1983) Stability of Infectious Pancreatic Necrosis Virus (IPNV) in untreated, filtered and autoclaved estuarine water. *Bulletin of the European Association of Fish Pathologists* **3**: 51–3.

Treasurer, J.W. and Cox, D. (1991) The occurrence of *Aeromonas salmonicida* in wrasse (Labridae) and implications for Atlantic salmon farming. *Bulletin of the European Association of Fish Pathologists* **11**: 208–10.

Treasurer, J.W. and Henderson, G. (1992) Wrasse: a new fishery. *World Fishing* **41**: 2.

Treasurer, J.W. and Laidler, A.L. (1994) *Aeromonas salmonicida* infection in wrasse (Labridae), used as cleaner fish, on an Atlantic salmon, *Salmo salar* L., farm. *Journal of Fish Diseases* **3**: 51–3.

Underwood, B.O., Smale, C.J., Brown, F. and Hill, B.J. (1977) Relationship of a virus from *Tellina tenuis* to Infectious Pancreatic Necrosis Virus. *Journal of General Virology* **36**: 93–109.

Wolf, K. (Ed.) (1988) *Fish Viruses and Fish Viral Diseases*. Cornell University Press, Ithaca, New York.

Chapter 20
Virulence of *Vibrio anguillarum* serotypes O1 and O2, *Aeromonas salmonicida* subsp. *salmonicida* and atypical *Aeromonas salmonicida* to goldsinny wrasse

K. GRAVNINGEN[1], P.G. KVENSETH[2] and R.O. HOVLID[3]

[1] *Apothekernes Laboratorium A.S., P.O. Box 158 Skøyen, N–0212 Oslo, Norway;* [2] *A/S MOWI, Bontelabo 2, P.O. Box 2, N–5023 Bergen, Norway; and* [3] *University of Bergen, Dept of Fisheries and Marine Biology, N–5023 Bergen, Norway*

Experience has shown that wrasse used as cleaner-fish in the production of Atlantic salmon can contract bacterial diseases such as vibriosis, and typical and atypical furunculosis, resulting in heavy wrasse mortality. *Vibrio anguillarum* serotypes O1 and O2 were shown to be highly virulent in experiments on goldsinny wrasse of *c.* 30 g.

20.1   Introduction

Wrasse used as cleaner-fish to remove ectoparasitic sea lice (Caligidae) from Atlantic salmon in net pens can develop bacterial diseases such as typical and atypical furunculosis (Collins *et al.*, 1991; Frerichs *et al.*, 1992). The diseases may cause heavy wrasse mortality (Treasurer & Laidler, 1994).

A project was started in May 1994, by A/S MOWI and Apothekernes Laboratorium A/S to investigate means of protecting wrasse against these diseases. As a preliminary study, the virulence of *V. anguillarum* serotype O1 and O2, *A. salmonicida* subsp. *salmonicida* and atypical *A. salmonicida* in wrasse were investigated.

20.2   Materials and methods

Goldsinny wrasse [*Ctenolabrus rupestris* (L.)] were obtained in April 1994 from fishermen in Grimstadfjorden, west of Bergen, Norway. The fish were acclimatised to water maintained at 11°C with a salinity of 34‰ and with a natural light regime in indoor tanks (2000 l). Goldsinny wrasse, of average weight 30 g and length 12 cm, were anaesthetized with benzocaine and challenged by different doses of bacteria by intraperitoneal injection (0.1 ml/fish). After inoculation the fish were transferred to 150 l tanks. The experiment consisted of two replicate tanks for each challenge isolate, each with 30 fish at two doses. The doses injected were 8.0

$\times 10^3$ and 8.0×10^5 bacteria/fish for *V. anguillarum* serotype O1 (AL 112) and *V. anguillarum* serotype O2 (AL 104), 3×10^2 and 3×10^4 bacteria/fish for *A. salmonicida* subsp. *salmonicida* (AL 2017) and 5×10^2 and 5×10^4 bacteria/fish for wrasse challenged to atypical *A. salmonicida* (AL 2058). All challenge isolates were clinical isolates from salmon, except the atypical *A. salmonicida* which was isolated from wrasse in Norway. The temperature during each challenge was 11–12°C, pH 8.0, $O_2 > 7$ mg/l and salinity 34‰.

Mortality was recorded daily. Necroscopy of 30% of the mortalities was performed on day seven for *V. anguillarum* serotype O1, day seven, nine and 13 for *V. anguillarum* serotype O2 and days 14, 15 and 22 for fish challenged to *A. salmonicida* subsp. *salmonicida* or atypical *A. salmonicida*. Bacteria were reisolated on blood agar and TYA from the kidney, and typed by serology.

20.3   Results

Goldsinny were highly susceptible to *V. anguillarum* serotype O1 and *V. anguillarum* serotype O2. The first mortality was recorded day four after inoculation, and cumulative mortalities after 21 days were 73.5 and 71% for serotype O1 and O2 respectively (low dose; 8×10^3) (Table 20.1). *V. anguillarum* serotype O1 or *V. anguillarum* serotype O2 were recovered from respective treatment tanks in all fish examined. LD_{50} for *V. anguillarum* serotype O1 and *V. anguillarum* serotype O2 were $<3 \times 10^3$ bacteria/goldsinny wrasse.

Mortality in goldsinny challenged to *A. salmonicida* or atypical *A. salmonicida* commenced between days five and ten, and the cumulative mortality after 28 days was similar in all tanks, independent of the dose and challenge isolate (73–97%) (Table 20.2). From all tanks, *A. salmonicida* subs. *salmonicida*, atypical *A. salmonicida* and *V. anguillarum* (not serotype O1 or O2) were recovered.

Table 20.1   Cumulative percentage mortality in goldsinny wrasse 21 days after intraperitoneal inoculation of two different doses of *V. anguillarum* serotype O1 or *V. anguillarum* serotype O2.

Challenge isolate	Bacteria/ fish	Replica No.	% mortality	Average % mortality	Bacteria recovered
V. anguillarum serotype O1	8×10^3	1	66	73.5	*V. anguillarum* serotype O1
		2	81		*V. anguillarum* serotype O1
V. anguillarum serotype O1	8×10^5	1	100	95	*V. anguillarum* serotype O1
		2	90		
V. anguillarum serotype O2	8×10^3	1	57	62.5	*V. anguillarum* serotype O2
		2	68		*V. anguillarum* serotype O2
V. anguillarum serotype O2	8×10^5	1	68	71	*V. anguillarum* serotype O2
		2	74		*V. anguillarum* serotype O2

Table 20.2 Cumulative percentage mortality in goldsinny wrasse 28 days after intraperitoneal inoculation of different doses of *A. salmonicida* subsp. *salmonicida* or atypical *A. salmonicida*.

Challenge isolate	Bacteria/ fish	Replica No.	% mortality	Average % mortality	Bacteria recovered
A. salmonicida subsp. *salmonicida*	3×10^2	1	73	83	*A. salmonicida* subsp. *salmonicida*, Atypical *A. salmonicida*, *V. anguillarum*[1]
		2	93		*A. salmonicida* subsp. *salmonicida*, Atypical *A. salmonicida*, *V. anguillarum*[1]
A. salmonicida subsp. *salmonicida*	3×10^4	1	81	81	*A. salmonicida* subsp. *salmonicida*, Atypical *A. salmonicida*, *V. anguillarum*[1]
		2	81		*A. salmonicida* subsp. *salmonicida*, Atypical *A. salmonicida*, *V. anguillarum*[1]
Atypical *A. salmonicida*	5×10^2	1	94	92	*A. salmonicida* subsp. *salmonicida*, Atypical *A. salmonicida*, *V. anguillarum*[1]
		2	90		*A. salmonicida* subsp. *salmonicida*, Atypical *A. salmonicida*, *V. anguillarum*[1]
Atypical *A. salmonicida*	5×10^4	1	91	94	*A. salmonicida* subsp. *salmonicida*, Atypical *A. salmonicida*, *V. anguillarum*[1]
		2	97		*A. salmonicida* subsp. *salmonicida*, Atypical *A. salmonicida*, *V. anguillarum*[1]

[1] Not *V. anguillarum* serotype O1 or serotype O2.

20.4   Discussion

Although the diseases mainly reported in wrasse used as cleaner-fish are atypical and typical *A. salmonicida* (Frerichs *et al.*, 1992; Collins *et al.*, 1991; Treasurer & Laider, 1994), *V. anguillarum* serotype O1 and *V. anguillarum* O2 are highly pathogenic to goldsinny. The virulence of *V. anguillarum* serotype O1 and *V. anguillarum* serotype O2 to goldsinny is similar to that reported by Toranzo *et al.* (1987) for turbot [*Scophthalmus maximus* (L.)], salmon (*Salmo salar* L.) and rainbow trout [*Oncorhynchus mykiss* (Walbaum)].

The recovery of different bacteria in fish challenged to *A. salmonicida* or atypical *A. salmonicida* might have been caused by latent carriers in the population. This makes it difficult to draw any conclusion regarding pathogenicity of these bacteria. However, atypical *A. salmonicida* was never diagnosed in goldsinny wrasse challenged to *V. anguillarum* serotype O1 or *V. anguillarum* serotype O2.

To give protection against bacterial diseases such as furunculosis, vibriosis and atypical furunculosis, vaccination of wrasse is recommended. Intensive cultured wrasse, unexposed to disease, will be of value in such experiments.

Acknowledgements

The work was carried out at the Industrial and Aquatic Laboratory in Bergen and was financially supported by NFR, The Norwegian Research Council.

References

Collins, R.O., Ferguson, D.A. and Bonniwell, M.A. (1991) Furunculosis in wrasse. *The Veterinary Record* **128**: 43.

Frerichs, G.N., Millar, S.D. and McManus, C. (1992) Atypical *Aeromonas salmonicida* isolated from healthy wrasse (*Ctenolabrus rupestris*). *Bulletin of the European Association of Fish Pathologists* **12**: 48–9.

Treasurer, J.W. and Laidler, L.A. (1994) *Aeromonas salmonicida* infection in wrasse (Labridae) used as cleaner-fish, on an Atlantic salmon, *Salmo salar* L., farm. *Journal of Fish Diseases* **17**: 155–61.

Toranzo, A.E., Santos, Y., Lemos, M.L., Ledo, A. and Bolinches, J. (1987) Homology of *Vibrio anguillarum* strains causing epizootics in turbot, salmon and trout reared on the Atlantic coast of Spain. *Aquaculture* **67**: 41–52.

Chapter 21
Determination of satiation and maintenance rations of goldsinny wrasse (*Ctenolabrus rupestris*) fed on mussel meat (*Mytilus edulis*)

D.B. GARFORTH[1], M.J. COSTELLO[2], R.D. FITZGERALD[1] and T.F. CROSS[1] [1] *Aquaculture Development Centre, University College Cork, Ireland and* [2] *Environmental Sciences Unit, Trinity College, Dublin, Ireland*

Goldsinny wrasse (*Ctenolabrus rupestris*) captured in the wild were held in a recirculating seawater facility. The influence of feeding frequency, a range of temperatures and body weight on daily food consumption of mussel meat (*Mytilus edulis*) was assessed. An overall mean satiation ration of 6% body weight per day was recorded for goldsinny of 14.7 g mean weight held at 16°C. Satiation ration consumed (expressed as a percentage of body weight) increased with increasing temperature and decreased with increasing fish body weight. The maintenance ration for goldsinny of similar mean weight (14.9 g) held at the same temperature, determined over an 80-day period was 4.13% bw/day. The application of these findings to the culture and maintenance of goldsinny used as cleaner-fish is discussed.

21.1   Introduction

A novel biological technique for controlling infestations of sea lice (Copepoda, Caligidae) on Atlantic salmon (*Salmo salar* L.) has been developed by using wrasse species (Teleostei, Labridae) found in the inshore waters of northern Europe as cleaner-fish. The goldsinny [*Ctenolabrus rupestris* (L.)] is the most commonly used cleaner-fish of sea lice (*Lepeophtheirus salmonis* Krøyer and *Caligus elongatus* Nordmann), which reflects both its abundance and widespread distribution (Costello *et al.*, *unpubl. data*).

The diet and growth of goldsinny in the wild and fish farm cages has been examined by Hildén (1978a, 1978b), Sayer *et al.* (1995) and Treasurer (1994), but little quantitative information exists of feeding and growth rates of the species in relation to temperature. Knowledge of maximum and maintenance rations may help maintain a healthy population of goldsinny in salmon cages, in the absence of sea lice and fouling organisms. This study aims to determine the daily satiation or maximum ration consumed by the goldsinny, by investigating the effects of feeding frequency, water temperature and body weight on consumption. The determined maximum ration for goldsinny of a known weight is then used to determine maintenance rations.

21.2   Materials and methods

Goldsinny wrasse were captured in black plastic shrimp pots from Casheen Bay, Co. Galway (Irish Nat. Grid Ref. L. 0842). They were held in a 5 m square fish cage until 104 specimens (body weight 9–40 g; length 85–106 mm) were brought to the Aquaculture Development Centre, University College, Cork in February, 1993. Fish were held in a recirculation facility under the same constant photoperiod (12 hours' light/12 hours' darkness). Fully oxygen-saturated seawater with 33–34.5‰ salinity was delivered at a flow rate of 700 l/h flow rate. The goldsinny were fed with mussel meat (*Mytilus edulis* L.). Experimentation commenced after a one month acclimatization period.

21.2.1   *Satiation ration determination*

Mussels were stored, uncooked, shells-on at −20°C. Small batches were defrosted at intervals of two to three days and then fed raw to the goldsinny. Satiation was achieved by offering a weighed amount of chopped mussel to each tank of fish until all feeding activity had ceased for two minutes. Uneaten mussel was then removed with a hand net and reweighed to obtain the total mussel consumed by the fish in each tank for each feed. In the event of feeding resuming during a two-minute period, more mussels were offered until a further full two minutes without any feeding activity occurring.

   In Experiment 1 a subpopulation of 64 wrasse were chosen (mean ± SD of 15.5 ± 2.05 g) and divided equally into four 500 l capacity circular tanks of 16 fish. A number of small sections of uPVC pipes (20–30 mm dia.) were provided in each tank to allow cover for the wrasse in order to reduce aggressive behaviour amongst individuals. Temperature was maintained at 14 ± 1°C throughout the trial. After 24 hours starvation, two tanks were fed to satiation three times a day and two tanks were fed to satiation twice a day during the twelve-hour light period for five days (Table 21.1). A similar experiment was then undertaken in Experiment 2 where goldsinny of mean ± SD weight 14.7 ± 2.68 g were held at 16°C ± 1°C, with feeding frequencies of seven and three satiation feeds per day. Feeding continued for three days, and this experiment was repeated twice, giving three replicates of Experiment 2 in total.

   To determine the effects of body weight on satiation, 15 each of small (mean ± SD; 9.6 ± 0.25 g), medium (19.0 ± 0.49 g) and large (32.0 ± 0.22 g) goldsinny were chosen for Experiment 3a and each fed to satiation three times a day for three days at a temperature of 13 ± 1°C. Experiment 3b was then repeated for 14 each of small (10.3 ± 1.35 g) and large (31.3 ± 4.66 g) goldsinny at 10 ± 1°C, fed in a similar fashion.

Table 21.1 Summary of satiation determination Experiments 1–3 for goldsinny of different mean weight held at different constant temperatures and fed at different frequencies in 500 l tanks (Experiment 2 was repeated three times).

Experiment	Mean weight (g)	Temp °C	No. of fish/tank	Feeds/day	Replicate tanks	Duration (days)
1	15.5 ± 2.05 (all tanks)	14 ± 1	16	×3 ×2	2 2	5
2	14.7 ± 2.68 (all tanks)	16 ± 1	16	×7 ×3	2 2	3
3a	9.6 ± 0.25	13 ± 1	15	×3	–	3
	19.0 ± 0.49	13 ± 1		×3	–	
	32.0 ± 0.22	13 ± 1		×3	–	
3b	10.3 ± 1.35	10 ± 1	14	×3	–	3
	31.3 ± 4.66	10 ± 1		×3	–	

Table 21.2 Percentage ration level of satiation, allocated to each pair of replicate tanks of 6 goldsinny, mean ± SD weight 14.9 ± 3.98 g held at 16 ± 1°C for an 80-day period.

% of satiation (7% bw/day)	% bw/day	Days fed 7% bw/day	Total days starved
100	7	80	–
80	5.6	64	16
60	4.2	48	32
40	2.8	32	48
20	1.4	16	64
0	0	0	80

21.2.2   *Maintenance ration determination*

To determine maintenance ration, 72 goldsinny of mean ± SD weight 14.9 ± 3.98 g were chosen and divided into 12 groups of six fish. Each group was placed in a 50 l white plastic circular tank, connected to the recirculation facility, and maintained at 16 ± 1°C. Again small sections of uPVC pipe were placed in each tank to provide hides for the wrasse.

All groups of fish were then fed for a two-week acclimatization period on a 4% bw/day ration; then random pairs of groups were allocated and fed a percentage of a predetermined satiation ration (7% bw/day) of chopped mussel meat (Table 21.2).

The goldsinny in all tanks were fed at the 7% bw/per day satiation level (derived from the three replicates of satiation Experiment 2). However, the various proportions of this ration allocated to each pair of tanks (0–100%) were achieved by starving the goldsinny on particular days: i.e. fish fed an 80% ration were fed at 7% bw for four days and starved the next, and so on for 80 days; fish fed a 60% ration were fed at 7% bw for three days and starved the next two etc.; and fish fed at 0% ration were starved throughout the 80-day period (Table 21.2). On each successive sixteenth day, all fish were starved and the weights of the six fish in each tank were recorded. The 100% ration (7%) was then adjusted according to the change in biomass of each tank over that 16-day period. The trial ran for 80 days.

All wrasse were anaesthetized in benzocaine (ethyl-p-amino benzoate) and blotted dry with a damp cloth prior to weighing. Goldsinny and mussels fed were weighed to 0.01 g with an electronic balance. Daily consumption and specific growth rate (SGR) in the maintenance determination were compared by unpaired Student's *t*-tests and analysis of variances (ANOVA) on a Statworks 1.2^{TM} computer package. Percentage body weight consumed (% bwc) was calculated as:

$$\% \text{ bwc} = \frac{\text{Mussel consumed}}{\text{Total weight of fish}} \times 100$$

SGR was calculated as:

$$\text{SGR} = \left(\frac{\text{Ln final weight} - \text{Ln initial weight}}{\text{Time period}}\right) \times 100$$

21.3   Results

21.3.1   *Satiation ration determination*

Mean ± SD daily consumption at 14°C ± 1°C was 3.8 ± 0.52 g and 4.12 ± 0.82 g for goldsinny fed two and three times per day, respectively. Three patterns were evident in the daily percentage bwc of each tank of goldsinny (Fig. 21.1): first percentage bwc of replicate tanks were similar ($P > 0.05$); second percentage bwc from different feed regimes on any day were similar ($P > 0.05$); and third consumption rate decreased from day one to three in both feed regimes. A significantly different amount was consumed between days one and five at each feed regime ($F_{(4,10)} = 7.489$); but on any single day there were no significant differences in consumption at either feed regime ($F_{(1,10)} = 2.561$) under test. However, when consumption on day one was eliminated from the analysis, there were no significant differences in satiation rations consumed between each of the remaining days, days two to five, for both feed regimes ($F_{(3,8)} = 7.813$). Mean ± SD daily consumption over the four days was 3.57 ± 0.28 g and 3.72 ± 0.22 g for the two- and three-feeds per day regimes, respectively.

The trend in consumption for all three replicate experiments of Experiment 2 for goldsinny held at 16°C were similar (Fig. 21.2) (replicate tanks combined in these analyses; $P > 0.05$). The first feed of the day was usually the greatest, and this showed most markedly on the first day. Comparing feeding frequencies, there was no significant difference in consumption per day for any single day ($F_{(1,6)}$; P

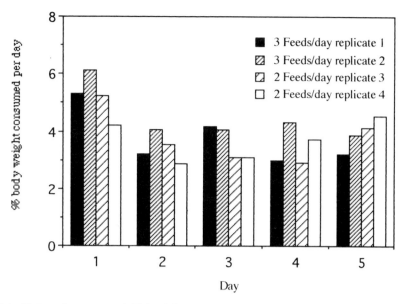

Fig. 21.1   Daily ration consumed (% bwc) for goldsinny fed two and three times per day.

> 0.05) between the seven- and three-feeds per day regimes for all three replicate experiments. Consumption on all days, at both feed frequencies, was similar in the first experiment ($F_{(1,4)}$ = 2.73). Consumption on day one, however, was significantly higher at both feed frequencies ($F_{(2,6)}$; $P < 0.05$) for the second and third replicate experiments. On elimination of day one, consumption on the remaining two days of the third replicate was similar ($F_{(1,4)}$; $P < 0.05$) but not so for the second replicate ($F_{(1-4)}$18.87; $P < 0.05$).

The overall means ± SD of daily satiation rate (% bwc) over the latter two days were 6.30 ± 1.11 g and 5.74 ± 0.87 g at the seven- and three-feeds per day regimes, respectively. A mean satiation ration of 6.02% bw/day for goldsinny of mean weight 15.0 g held at 16 ± 1°C was obtained from pooling the two-day mean values from both the seven- and three-feeds per day regime.

Maximum daily ration decreased significantly with increasing fish size when daily consumption rate of each group of goldsinny in Experiment 3a was plotted ($F_{(2,4)}$ = 62.52; $P < 0.001$) (Fig. 21.3). Mean ± SD daily rations were 4.44 ± 0.25 g, 3.59 ± 0.49 g, and 1.94 ± 0.22 g for small, medium and large goldsinny, respectively. Again, small goldsinny consumed more food, expressed as a percentage of body weight per day, than larger fish in Experiment 3b (Fig. 21.4). However, this difference was not significant ($F_{(1,2)}$ = 3.85; $P > 0.05$).

An exponential regression line for all size groups of goldsinny at the different temperatures under which they were fed in Experiments 1 to 3 (Fig. 21.5) revealed body weight to be negatively correlated to satiation ration % bwc = $6.705 \times 10^{-0.015w}$: d.f. = 5, r = 0.834; $P < 0.05$).

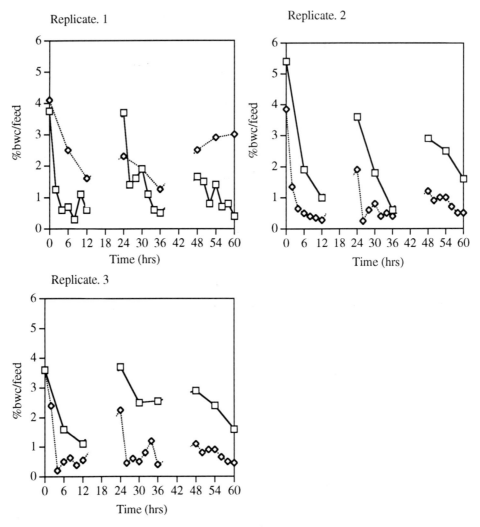

Fig. 21.2 Influence of feeding frequency on satiation ration consumed per meal for goldsinny fed three and seven times per day for three replicate experiments. Each symbol relates to one feed.

21.3.2   *Maintenance ration determination*

Mean weight increased gradually for all goldsinny fed above the 40% ration level at 16 ± 1°C (Fig. 21.6). At the 40% ration level, mean weight decreased after the first 16 days. Goldsinny lost weight on ration levels of 40% or below. Specific growth rate (SGR) was plotted over each 16-day period at each ration level (Fig. 21.7). SGRs were highly significant between rations ($F_{(5,20)}$ = 8.263; $P < 0.001$) and over the experimental time ($F_{(4,20)}$ = 3.270; $P < 0.05$).

When SGR equals zero the corresponding ration is the maintenance proportion of the total ration consumed. This became an increasingly larger proportion of the

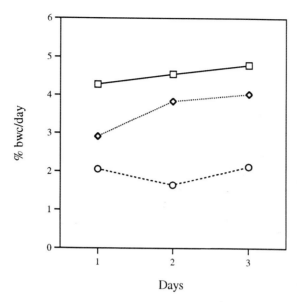

Fig. 21.3 Daily ration consumed (% bwc/day) for three size classes of goldsinny; 9.6 g (squares); 19.0 g (diamonds); and 31.1 g (circles), held at 13 ± 1°C.

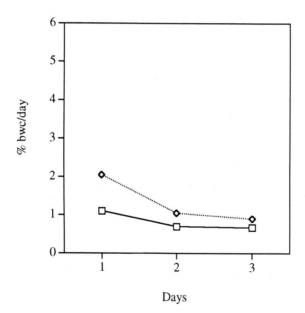

Fig. 21.4 Daily ration consumed (% bwc/day) of two size classes of goldsinny: 10.3 g (squares) and 31.1 g (diamonds) held at 10 ± 1°C.

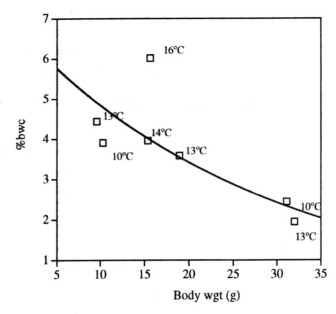

Fig. 21.5   Daily ration consumed for each size class at each temperature.

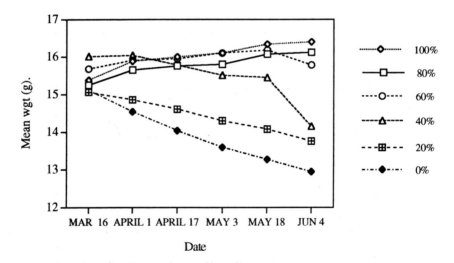

Fig. 21.6   Mean weight (actual growth) of goldsinny fed each ration, measured in 16-day periods over 80 days.

maximum 7% bw/day ration over each successive 16-day experimental period (2.59%, 3.92%, 4.06%, 3.64%, 3.78% bw/day). The overall SGR for the entire experimental period (Fig. 21.8) at each ration gave a maintenance ration of 4.13% of goldsinny body weight per day, equivalent to 59% of the maximum ration fed.

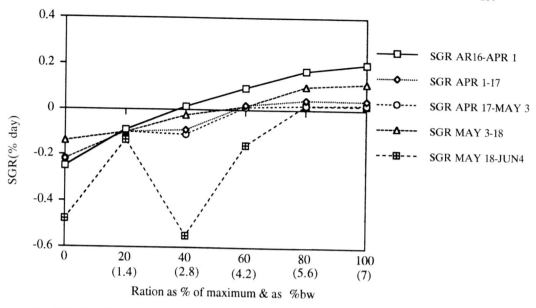

Fig. 21.7 Specific growth rate (SGR) over each 16-day period for goldsinny fed 0–100% of satiation ration.

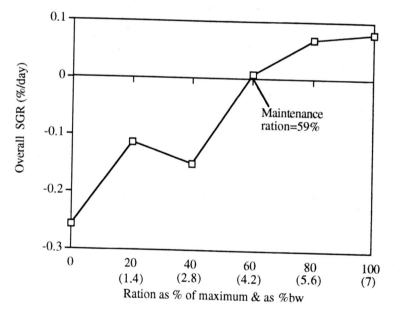

Fig. 21.8 Specific growth rate (SGR) over the entire experimental period at 16 ± 1°C showing the maintenance ration for the 80-day period.

21.4   Discussion

21.4.1   *Influence of feeding frequency on satiation ration*

There were no significant differences in mean daily rations when feeding the goldsinny at frequencies of two, three and seven times per day. This is similar to salmonids, where maximum daily food intake of rainbow trout [*Oncorhynchus mykiss* (Walbaum)] was achieved with just two satiation feeds per day (Grayton & Beamish, 1977). There was no difference in mean daily ration consumed when fed either two, three or six meals per day, although significantly less was eaten when only one satiation feed per day was given. Similarly, sea-bass fingerlings [*Dicentrarchus labrax* (L.)] had almost the same daily rate of feeding when fed either three or four times daily (Tsevis *et al.*, 1992). Feeding frequency, producing maximum growth, was achieved on two feeds per day in the case of these two studies. However, meal frequencies of less than one per day resulted in weight loss in winter flounder [*Pseudopleuronectes americanus* (Walbaum)] (Tyler & Dunn, 1976).

Studies on the effects of the feeding frequency of satiation ration on growth rate of the goldsinny over an extended period was not the intention of this study. Rather, frequency of feeding was investigated to test how infrequently the goldsinny could be fed in order to meet their mean daily satiation ration. On the basis of the present experiments, it was concluded that two feeds per day were sufficient.

As goldsinny were starved for 24 hours prior to the onset of the trial, it is reasonable to assume that the fish had overcompensated their feed requirements on the first day due to the previous 24-hr starvation period. In most cases, when the first day was eliminated, daily consumption was similar on remaining days. Periods of short starvation, which allow the gut to digest, absorb and evacuate previously ingested material, to some extent govern the degree of gorging at subsequent meals (Grove *et al.*, 1978). The 12-hr dark period and 24-hr food deprivation periods within the present trials would allow for different degrees of evacuation to take place. However, the various fasting periods in between each satiation meal of Experiment 1 would allow for much less clearance time of the stomach and gut. Subsequently, this influenced consumption at each satiation meal.

21.4.2   *Influence of temperature on satiation ration*

Satiation ration of the goldsinny increased with temperature, as in other fish (Elliott, 1975; Grove *et al.*, 1978; Jobling, 1988). This relationship is probably true over the temperature range in this study. However, the overall relationship is probably more complex at temperatures beyond the tolerance of the goldsinny. For example, at temperatures of 10°C or less, goldsinny are known to enter a torpid state where feeding activity is greatly diminished (Costello, 1991; Darwall *et al.*, 1992). Ration consumed would constitute that of maintenance rather than for growth purposes. In addition, gut evacuation increases with increasing temperature (Windell *et al.*, 1976; Grove *et al.*, 1985) and since stomach emptiness has an influence on the ration consumed at each meal, it is likely that the relationship of temperature on gut evacuation is also involved in the increase in satiation ration with temperature.

21.4.3   *Influence of body weight on satiation ration*

Generally, satiation ration expressed as percentage body weight decreases with increased body weight (Brett, 1979; Elliott, 1975; Austreng *et al.*, 1987). This was observed in Experiments 3a and 3b. In addition, the relationship of body weight (w) on satiation ration consumed (% bw/day) was exponential; % bwc = 6.705 × $10^{-0.015w}$ ($r = 0.833$; $P < 0.05$), though this relationship may be more complex for very small and large goldsinny outside the limits investigated.

There is a gradual decrease in growth rate of animals with increasing size (Brody, 1945; von Bertalanffy, 1957), since growth hormone is most active in young, rapidly growing fish, and energy requirements for standard metabolism are higher (Brett and Groves, 1979). Therefore, the smaller, younger goldsinny consumed a larger ration (expressed as % bw) than larger fish.

21.4.4   *Maintenance ration determination*

A feeding hierarchy within each tank was a likely occurence. Temporal variation in the feeding of a maximum ration (100%) at varying time intervals (days) proved to be an effective methodology by which all goldsinny would have an increased opportunity to consume some proportion of the mussel meats offered. The 100% ration used was taken as 7% bw/day. This was slightly above the observed satiation ration determined for similar sized fish held at 16 ± 1°C in Experiment 2 (6.02%). By feeding slightly above the determined ration, it was envisaged that satiation was always achieved on the 100% ration level.

The relationship of increased ration promoting increased growth rate has been well documented (Staples and Nomura, 1976; Brett, 1979; Houlihan *et al.*, 1988). By the end of the experimental period only fish fed on an 80% ration or above showed an increased biomass (Fig. 21.6). On the 60% ration, growth had been recorded previous to the final 16-day interval, albeit at a very small rate. In the same way, fish fed on a 40% ration initially showed a very small increase in mean weight. On reduced rations and in starving fish, any available fat reserves present were probably expended to supplement the energy derived from the ration for the purposes of basal metabolism; as these became diminished, muscle protein was probably utilized for energy and amino acid requirements (Flowerdew & Grove, 1980). An increase in water content with loss of lipid and protein has been shown to compensate overall weight loss (Staples & Nomura, 1976; Flowerdew & Grove, 1980). This may have been the case for goldsinny on reduced rations.

Specific growth rates obtained on the various rations were significantly different over the entire experimental period. Specific growth over the whole period was highest on the larger ration levels in agreement with the bulk of work on fish (Elliott, 1975; Brett & Groves, 1979; Houlihan *et al.*, 1988; Houlihan *et al.*, 1989). However, growth on the 100% ration and all other rations declined throughout the experimental period. Goldsinny fed a 60% ration, for example, began to lose weight before the end of the experiment.

Since body weight and temperature influence maintenance requirements (Elliott, 1976), the maintenance ration determined here, of 4.13% bw/day over the entire experimental period, applied only to goldsinny of that body weight range held at the temperature under investigation (16 ± 1°C).

The maintenance ration increased from 2.59% bw/day to 3.92% bw/day over the first 16-day period, and thereafter remained fairly constant until a final figure of 5.46% bw/day of the maximum satiation ration (Fig. 21.7). Maintenance ration increased linearly with fish weight for thick-lipped grey mullet [*Chelon (= Crenimugil) labrosus* (Risso)] (Flowerdew & Grove, 1980), and this was attributed to uncontrolled differences in spontaneous activity. The increase in goldsinny weight over the experimental period would result in an associated requirement for energy. However, rate of standard metabolism usually falls with increasing body weight when expressed as a proportion (%) of weight (Brett & Groves, 1979). Although more energy was required to maintain the goldsinny as they grew, it is reasonable to assume that maintenance requirement per unit weight increment would remain fairly constant over the experimental period and ultimately decline with increasing fish size and age.

The large initial shift in maintenance requirement (2.59% to 3.92% bw/day) may have been attributable to several factors. Fish energy stores (lipid) were more abundant in all groups at the start of the experiment and these, in addition to whatever energy was derived from the actual ration fed (at or above the 40% level) would be sufficient for both maintenance and growth purposes during the early stages. This is reflected in the initial positive growth, albeit extremely slight, of goldsinny fed on a 40% ration. Later, as lipid energy stores were diminished, energy for both maintenance and growth had to be derived solely from the rations fed and thus an increased proportion of the ration was required for maintenance.

Alternatively, tank conditions may have created stressful situations through competition for space and food which may have contributed to abnormally high maintenance requirements throughout the experiment. Aggressive behaviour and damage caused to individuals may have also contributed to excessive maintenance requirements. In addition, at the 60% ration level, four mortalities occurred during the last 16-day period which probably resulted in a spuriously high maintenance requirement figure of 5.46% bw/day. Finally, since the experiment was carried out at the beginning of the natural spawning period, maturation of some individuals may have occurred to some extent. This again, may have increased maintenance requirements.

21.4.5   Implications for aquaculture

The estimations of maximum and maintenance rations obtained throughout the study and the growth-rates observed on the varying rations have a twofold relevance to aquaculture. If the culture of the goldsinny wrasse is to take place commercially, rather than on a scientific basis, information of maximum daily food intake under varying ambient conditions is essential so that growth rates may be maximized and wrasse of a viable stocking size produced in less time than in the wild. A healthy population of fish of a large enough size would be the prime objective when considering for stocking within cages to de-louse salmon. This combined with increased survival in captivity, more nutritional information on dietary requirements and the assurance of disease-free wrasse for stocking with salmon may prove wrasse culture an economically viable venture in the near future. Wild caught juveniles could also be ongrown in small meshed net pens for later stocking with salmon, their growth rates optimized by regular feeding of a maximum ration.

Although mussel meats were deemed a suitable feed source for goldsinny, analysis of the nutritional status of various diets would be desirable in the formulation of wrasse diets.

The maintenance ration estimation for the goldsinny may prove useful in terms of maintaining a population of fish within the salmon cage, when there is an absence of other food material, i.e. lice or other prey items. This would be most apparent after fouled nets are changed for clean ones. Feed delivery to wrasse may prove difficult in large salmon cages, although the authors have successfully hung the mussel meats in mesh bags from the sides of the cage net, thereby allowing accessibility to the goldsinny within.

Acknowledgements

This study was supported by contract AQ.2.502 on Cleaner-fish Technology from the Fisheries and Aquaculture Research Programme of the European Community.

References

Austreng, E., Storebakken, T. and Asgard, T. (1987) Growth rate estimates for cultured Atlantic salmon and rainbow trout. *Aquaculture* **60**: 157–60.

Brett, J.R. (1979) Environmental factors and growth. In *Fish Physiology* Vol. VIII (Ed. by Hoar, W.S., Randall, D.J. and Brett, J.R.). Academic Press, London, pp. 599–675.

Brett, J.R. and Groves, T.D.D. (1979) Physiological energetics. In *Fish Physiology* Vol. VIII (Ed. by Hoar, W.S., Randall, D.J. and Brett, J.R.). Academic Press, London, pp. 279–352.

Brody, S. (1945) *Bioenergetics and Growth*. Reinhold, New York, pp. 470–565.

Costello, M.J. (1991) Review of the biology of wrasse (Labridae: Pisces) in Northern Europe. *Progress in Underwater Science* **16**: 29–51.

Darwall, W.R.T., Costello, M.J., Donnelly, R. and Lysaght, S. (1992) Implications of life-history strategies for a new wrasse fishery. *Journal of Fish Biology* **41** (Supplement B): 111–23.

Elliott, J.M. (1975) Weight of food and time required to satiate brown trout, *Salmo trutta* L. *Freshwater Biology* **5**: 51–64.

Elliott, J.M. (1976) The energetics of feeding, metabolism and growth of brown trout (*Salmo trutta* L.) in relation to body weight, water temperature and ration size. *Journal of Animal Ecology* **45**: 923–48.

Flowerdew, M.W. and Grove, D.J. (1980) An energy budget for juvenile thick-lipped mullet, *Crenimugil labrosus* (Risso). *Journal of Fish Biology* **17**: 395–410.

Grayton, B.D. and Beamish, F.W.H. (1977) Effects of feeding frequency on food intake, growth and body composition of rainbow trout (*Salmo gairdneri*). *Aquaculture* **11**: 159–72.

Grove, D.J., Loizides, L.G. and Nott, J. (1978) Satiation amount, frequency of feeding and gastric emptying rate in *Salmo gairdneri*. *Journal of Fish Biology* **12**: 507–16.

Grove, D.J., Moctezuma, M.A., Flett, H.R.J., Foott, J.S., Watson, T. and Flowerdew, M.W. (1985) Gastric emptying and the return of appetite in juvenile turbot, *Scophthalmus maximus* L., fed on artificial diets. *Journal of Fish Biology* **26**: 339–54.

Hilldén, N.O. (1978a) On the feeding of the goldsinny, *Ctenolabrus rupestris* L. (Pisces, Labridae). *Ophelia* **17**: 195–8.

Hilldén, N.O. (1978b) An age-length key for *Ctenolabrus rupestris* (L.) in Swedish waters. *Journal du Conseil International pour l'Exploration de la Mer* **38**: 270–71.

Houlihan, D.F., Hall, S.J., Gray, C. and Noble, B.S. (1988) Growth rates and protein turnover in Atlantic cod, *Gadus morhua*. *Canadian Journal of Fisheries and Aquatic Sciences* **45**: 951–64.

Houlihan, D.F., Hall, S.J. and Gray, C. (1989) Effects of ration on protein turnover in cod. *Aquaculture* **79**: 103–10.

Jobling, M. (1988) A review of the physiological and nutritional energetics of cod, *Gadus morhua* L., with particular reference to growth under farmed conditions. *Aquaculture* **70**: 1–19.

Sayer, M.D.J., Gibson, R.N. and Atkinson, R.J.A. (1995) Growth, diet and condition of goldsinny, *Ctenolabrus rupestris* (L.) on the west coast of Scotland. *Journal of Fish Biology* **46**: 317–40.

Staples, D.J. and Nomura, M. (1976) Influence of body size and food ration on the energy budget of rainbow trout *Salmo gairdneri* Richardson. *Journal of Fish Biology* **9**: 29–43.

Treasurer, J.W. (1994) Prey selection and daily food consumption by a cleaner fish, *Ctenolabrus rupestris* (L.), on farmed Atlantic salmon, *Salmo salar* L. *Aquaculture* **122**: 269–77.

Tsevis, N., Klaoudatos, S. and Conides, A. (1992) Food conversion budget in sea bass, *Dicentrarchus labrax*, fingerlings under two different feeding frequency patterns. *Aquaculture* **101**: 293–304.

Tyler, A.V. and Dunn, R.S. (1976) Ration, growth and measure of somatic and organ condition in relation to meal frequency in winter flounder, *Pseudopleuronectes americanus*, with hypotheses regarding population homeostasis. *Journal of the Fisheries Research Board of Canada* **33**: 63–75.

von Bertalanffy, L. (1957) Quantitative laws in metabolism and growth. *Quarterly Review of Biology* **32**: 217–31.

Windell, J.T., Kitchell, J.F., Norris, D.O., Norris, J.S. and Foltz, J.W. (1976) Temperature and rate of gastric evacuation by rainbow trout, *Salmo gairdneri*. *Transactions of the American Fisheries Society* **6**: 712–17.

Chapter 22
Successful survival of wrasse through winter in submersible netcages in a fjord in western Norway

R. BJELLAND[1], J. LØKKE[1], L. SIMENSEN[1] and P.G. KVENSETH[2]

[1] *Sogn and Fjordane College, P.O. Box 133, N–5801 Sogndal, Norway; and* [2] *A/S Mowi, Bontelabo 2, P.O. Box 4102, Dreggen, N–5023 Bergen, Norway*

Most wrasse stocked with salmon die during the winter. A trial was undertaken to try to improve winter survival using four closed net cages ($3\,m^3$) each filled with approximately 200 wrasse. Two cages were deployed in late October 1993 at a depth of 17 m and two at 25 m at Åkre in Hardangerfjord in western Norway. At each depth, one cage was filled with goldsinny [*Ctenolabrus rupestris* (L.); total length 8–14 cm], and the other with corkwing [*Crenilabrus melops* (L.); total length 9–18 cm]. The cages were inspected monthly. In late April 1994, the cages were raised to the surface. An average of 94% goldsinny survived, compared with only 13% corkwing. The average condition factor (CF) decreased for all groups during the experiment, and was most pronounced for corkwing. Deployment at different depths did not influence the decrease in CF.

22.1   Introduction

In commercial Atlantic salmon (*Salmo salar* L.) farming in Norwegian waters, smolts are attacked by sea lice (*Lepeophtheirus salmonis* Krøyer and *Caligus elongatus* Nordmann) immediately after release to seawater from April to May (Viga, *pers. comm.*). Where wrasse are used to control sea lice infestations, few are available at the time of first stocking because of low catches to the fishery. In addition, most wrasse kept with salmon in net pens die during winter (Kvenseth, 1993). Successful maintenance of wrasse through winter would make them available before the first sea lice infestations. Improved survival of wrasse over winter would also be important in avoiding possible local over-fishing by reducing annual demand. Preliminary experiments on overwintering wrasse in shallow water (2–13 m) gave promising results with 34–68% survival (Skog *et al.*, 1994). In nature, some wrasse may migrate to deeper waters in winter (Hilldén, 1984). The aim of the present study was to increase winter survival of wrasse by deploying cages at greater depths than those used by Skog *et al.* (1994).

20.2   Materials and methods

The experiment was performed at a MOWI salmon farm at Åkre in the Hardangerfjord in Western Norway, from 28 October 1993 until 20 April 1994.

Wrasse were caught by fyke nets within a range of 1 km from the salmon farm and stored for about one to two weeks in an empty pen prior to the experiment. Only wrasse without wounds or signs of disease were used in the experiment. Wrasse were sorted by species into two groups: 358 goldsinny [*Ctenolabrus rupestris* (L.)] and 369 corkwing [*Crenilabrus melops* (L.)].

Wrasse were maintained in submersible cages at two different depths, 17 and 25 m. The species groups were divided between two cages at each depth. The species were kept in separate cages to avoid any interspecific competition. The cages measured 1 × 1 × 3 m and were made with a solid wooden floor and a wooden frame covered with 12 × 12 mm² net (Fig. 22.1).

Shelters for the wrasse were provided in each cage, made of sectioned (40 cm long) plastic pipes (50 mm dia.), tied together and attached to the floor. There were 80 pipes in each cage. In addition, two black plastic bags were shredded to look and function as artificial seaweed. Each cage was anchored to 100 kg concrete blocks resting on the fjord bottom (Fig. 22.1). For stabilization small

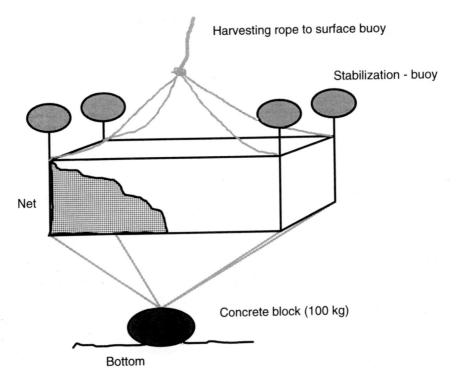

Fig. 22.1 Submersible net cage used to keep goldsinny and corkwing wrasse through the winter.

subsurface buoys were located at each corner of the cages. A harvesting rope was attatched to the top of each cage. The cages were not influenced by wind or surface wave action.

Before the cages were submerged, about 45% of the fish were measured at random for total length (TL) to an accuracy of 1 mm, and wet weight (W) to the nearest g. A condition factor (CF) was calculated using Fulton's formula:

$$(W \times 100)/TL^3,$$

where W was in g, TL in cm. The fish were not anaesthetized before measuring. A sample of 30 goldsinny and 30 corkwing were killed and examined immediately for live ectoparasites.

The cages filled with wrasse were quickly lowered to the nominal depths. To avoid stressing the fish while submerged, they were not fed and no samples were taken during the experimental period. The cages were inspected monthly, except in January, by SCUBA divers, and the activities of the wrasse recorded for about 30 min. by video camera.

Temperature and salinity recordings were made daily at depths of 1 and 10 m. Measurements were only sporadically performed at 20 m from November to April. Temperature was measured with an accuracy of 0.1°C using a mercury thermometer. Salinity was estimated by means of a densimeter to the nearest ‰.

At the end of the experiment, the cages were raised to the surface over time periods of 5–30 min. All surviving fish were transferred to tanks with fresh seawater, and length and weight measured for all living corkwing, and 60% of living goldsinny. A random selection of about 20 surviving fish from each cage were examined for ectoparasites.

22.3   Results and discussion

22.3.1   *Water quality and wrasse activity*

The lowest temperature during the experimental period at 10 m depth was 2.8°C (27 February 1994), the highest 12.1°C (1, 3 and 11 October 1993) (Fig. 22.2). The lowest temperature at 20 m was 3.9°C (21 March 1994), the highest 10.2°C (8 November 1993).

The video recordings showed decreasing activity of wrasse through winter, although this was neither defined or quantified. It was assumed that goldsinny were in a torpid state in periods from January to the end of the experiment. This state of torpor has previously been reported for goldsinny (Hilldén, 1984; Kvenseth, 1993; Chapter 10, this volume); corkwing have not been observed in this condition (Hilldén, 1984). During January to April the average temperature at depths of 10 and 20 m was below 5°C, in the range 2.8–5.5°C at 10 m, and 3.9–5.3°C at 20 m.

Throughout the experimental period, salinity was highest at 20 m depth (Fig.

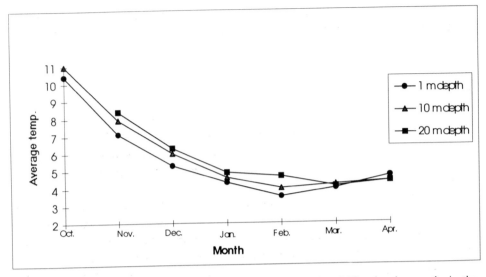

Fig. 22.2   Average monthly temperature (°C) at depths of 1, 10 and 20 m for the months in the experimental period.

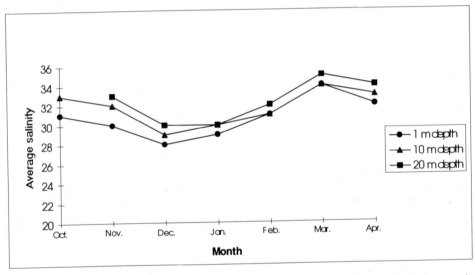

Fig. 22.3   Average monthly salinity (‰) at depths of 1, 10 and 20 m for the months in the experimental period.

22.3). The lowest salinity measured at 10 m depth was 24‰ (December and January), and the highest 35‰ (October, November, January, March and April). The lowest salinity at 20 m was 25‰ (December and January), and the highest 35‰ (November, December, February, March and April).

According to Sayer *et al.* (this volume, Chapter 10) there is little evidence of vulnerability to low salinity for either corkwing or goldsinny. As the lowest salinity measurement during the experiment was 25‰, it is assumed that salinity had no influence on mortality.

22.3.2   *Wrasse survival*

There were marked differences in survival between the species during the six months of submergence (Table 22.1). A total of 91.8% of goldsinny survived at 17 m depths, and 95.5% at 25 m. For corkwing, 12.4% survived at 17 m, 14.0% at 25 m. For both species, there were no significant differences between the two depths (Fishers exact test, $\alpha = 0.05$).

If goldsinny were in a torpid state for most of the experimental period, this might explain the higher level of survival. According to Hilldén (1984), goldsinny do not eat at temperatures below 5°C. Lack of food would therefore be unlikely to affect survival. The relatively high densities at which they were maintained did not appear to affect goldsinny survival. Goldsinny have been observed crowded in crevices during winter (Harberg, *pers. comm.*), but the natural behaviour of corkwing is not well known. High corkwing mortality in this experiment indicates that they may have other winter survival strategies in the field compared with goldsinny.

22.3.3   *Wrasse condition*

The average weight of corkwing decreased during the experimental period (Table 22.2). The largest reduction was observed for corkwing at 25 m, with an average loss of 23% whole body weight. The average whole body weight reduction of corkwing at 17 m depth was 11%. Average length decreased for both groups of corkwing (Table 22.2). Corkwing CF decreased from 1.6 to 1.3 at both depths. Because of high corkwing mortality, it is difficult to draw any firm conclusions from these results, but the decrease in CF may reflect poor living conditions.

Goldsinny gained weight during the over-wintering by an average of 5.6% body weight at both depths. The average length of goldsinny increased at both depths.

Table 22.1 Survival of corkwing and goldsinny at 17 and 25 m depths.

Species	Depth (m)	Number at start	Number of survivors	Dead counted	Survival (%)
Corkwing	17	169	21	14	12.4
	25	200	28	7	14.0
Goldsinny	17	158	145	2	91.8
	25	200	191	0	95.5

Table 22.2   Weight (W), total length (L) and condition factor (CF) of corkwing and goldsinny before and after submergence.

	Species	Depth (m)	Number measured	Mean W (g) (range)	Mean L (cm) (range)	Mean CF (range)
Before submergence	Corkwing		169	35 (11–96)	12.8 (8.7–18.4)	1.6 (1.0–2.9)
	Goldsinny		158	18 (9–47)	10.6 (8.3–14.1)	1.4 (0.8–2.8)
After submergence	Corkwing	17	21	31 (14–56)	13.0 (11.1–15.8)	1.3 (1.0–1.8)
	Goldsinny	17	86	19 (10–47)	11.3 (9.4–14.5)	1.3 (0.9–1.6)
	Corkwing	25	28	27 (10–64)	12.5 (9.2–16.5)	1.3 (1.0–1.5)
	Goldsinny	25	120	19 (10–38)	11.1 (9.0–13.9)	1.3 (0.7–1.7)

However, both these results may be explained by a disproportionate level of mortality among smaller fish. Goldsinny CF decreased from 1.4 to 1.3 (Table 22.2).

22.3.4   *Ectoparasitic infection*

Surviving corkwing were considerably more infected with ectoparasites, mainly *Trichodina* spp. Ehrenberg, than goldsinny.

22.4   **Conclusions**

The experiment demonstrated the possibility of storing goldsinny successfully through winter in submersible netcages with high levels of survival. Corkwing were not as tolerant to maintenance under similar conditions. The behaviour of goldsinny at low temperatures, where they are observed to enter a state of torpor, may be an adaption which enhances survival over winter. These results accord with the work of Sayer *et al.* (Chapter 10, this volume) on the physiology of wrasse.

Acknowledgements

We are grateful to the staff at Åkre fish farm for helpful assistance during the experiments. We are also grateful to Anne Mette Kvenseth and Torbjørn Dale, for suggesting improvements to the manuscript. This study was supported financially by NRF, the Norwegian Research Council.

References

Hilldén, N.O. (1984) Daily, seasonal and group behaviour of the goldsinny wrasse, *Ctenolabrus rupestris*, (*Teleostei, Labridae*) in the Swedish Skagerrak, Tjärnö Marine Biological Laboratory. Unpublished Doctoral thesis, University of Stockholm, Stockholm.

Kvenseth, P.G. (1993) Use of wrasse to control salmon lice. In *Fish Farming Technology* (Ed. by Reinertsen, H., Dahle, L.A., Jørgensen, L. and Tvinereim, K.). Balkema, Rotterdam, pp. 227–32.

Skog, K., Mikkelsen, K.O. and Bjordal, Å. (1994) Overvintring av bergnebb (*Ctenolabrus rupestris*) i lukkede teiner. *Fisken og havet* **5**: 1–5.

Chapter 23
Wrasse biology and aquaculture applications: commentary and conclusions

M.D.J. SAYER[1], J.W. TREASURER[2] and M.J. COSTELLO[3] [1] *Centre for Coastal and Marine Sciences, Dunstaffnage Marine Laboratory, P.O. Box 3, Oban, Argyll PA34 4AD, UK;* [2] *Marine Harvest McConnell, Lochailort, Inverness-shire PH38 4LZ, UK; and* [3] *Environmental Sciences Unit, Trinity College, Dublin 2, Ireland*

Most of the work carried out on wrasse over the past five to seven years is either reported on directly in this volume, or summarized in review chapters. Because this is the first publication dedicated exclusively to wrasse biology and applications for aquaculture, it is thought appropriate that a commentary on the contents should be made, in order to highlight areas of future research and the possible expansion of the technique to other aquaculture industries.

The use of north European wrasse species as cleaner-fish in the Atlantic salmon culture industry has proved a worthy alternative to pesticides (this volume, Chapters 13, 14, 15, 16). Not only are wrasse effective in sea lice control (Chapter 14) but they can also be employed to clean salmon nets of fouling organisms (Chapter 15). The interest in wrasse shown by the salmon farming industry has resulted in a number of biological studies being undertaken on a family of fishes largely ignored prior to the mid to late 1980s (Chapters 2–10).

A positive feature to the use of wrasse in aquaculture has been the willingness of the salmon farming industry to fund fundamental biological research from a very early stage in the adoption of the technique. There were obvious benefits to the industry in knowing, for example, the age structure of wrasse, its population sustainability and potential disease problems. However, this should not detract from the large contribution to our knowledge of the biology of north European wrasse that has been made in the past few years in particular. Many aspects of the biology and behaviour of goldsinny are now well documented (see Chapters 2, 3, 5, 8, this volume), but details are still lacking in some areas for corkwing and rock cook, wrasse species which are both commercially exploited. The geographical variation discovered for some biological parameters (Chapter 2) has implications for wrasse fisheries that may, in future, be widespread anywhere from the southern Mediterranean to mid-Norway. Observations of cleaning made in the field (Chapter 4) were similar to those which first attracted fish biologists to wrasse as potential cleaner-species. Similar field studies which detail any temporal, spatial or behavioural differences in cleaning, territorial or reproductive activity in the field (Chapters 3 and 5) may influence husbandry requirements for wrasse in salmon cages.

The total reliance on traps as methods of wrasse capture, makes the wrasse fishery distinctive in north European waters and worthy of study (Chapters 6 and 7). Although not a lot of detailed examination has been made of the selectivity of the fishing techniques employed, catch records have yielded valuable information on species distribution. Most importantly, a detailed assessment of wrasse presence was made close to the northern limits of distribution (Chapter 6). Concerns as to the sustainability of local wrasse populations were raised early during the onset of exploitation, prompting studies of the ecology of the fisheries (Chapters 8 and 9). The complexity of the biology of inshore wrasse populations, and the habitat in which they are fished, made any confident predictions of future fishery status unlikely. Although caution was recommended in the chapters which dealt with wrasse fisheries (Chapters 8 and 9), to some extent the fisheries have the potential to be self-regulating because of the minimum size limitations imposed by the industry on commercial suppliers (Chapter 7, also Appendix 1).

It is evident, from some of the work presented in this volume, that further improvements in wrasse husbandry can be made. Discussions held at the symposium on which these proceedings are based resulted in a set of guidelines for the use of wrasse on farms being produced (Appendix 1). High escapement and losses over winter are obstacles to extending the success of the technique into the second year of the salmon production cycle. Where the same wrasse have been kept with salmon from first introduction to harvest, effective cleaning has been recorded, but introducing wrasse with salmon of about 1 kg has been unsuccessful in some trials (Chapter 16). Following the guidelines in Appendix 1 may result in improved winter survival of wrasse, and improved cleaning efficiency during the second year that the salmon are at sea. Cleaning efficiency may improve if nets are kept clean (Chapter 16; Appendix 1) but this may necessitate dietary augmentation for wrasse during periods when sea lice numbers are low (Chapter 21). Winter wrasse mortality appears to be affected by salinity variations in association with low temperatures, although there are interspecific differences, with goldsinny apparently more robust than either corkwing or rock cook (Chapter 10). Two options available for improving wrasse survival are overwintering in submersible cages at depths below 17 metres (Chapter 22), or the provision of improved shelters in salmon cages.

There are genuine fears in the salmon farming industry concerning potential transfer of pathogens from wild-caught wrasse to salmon in cages. The impact of infectious pancreatic necrosis, pancreas disease, and other viral agents of wrasse appears not to be of major concern (Chapter 19). Similarly, the parasite fauna of wrasse should not pose a significant risk to salmon (Chapters 17 and 18). However, vibriosis and *Aeromonas salmonicida* have been recorded in wrasse (Chapter 20), the latter in both typical and atypical strains. Atypical furunculosis is common in several wild marine fishes and its pathogenicity to salmon has still to be determined. These findings indicate the importance of vaccinating wrasse against furunculosis, of screening wrasse from new sources, and of obtaining wrasse locally but at a distance from other farms where possible (Appendix 1).

The potential for exporting cleaner-fish technology to other parts of the world and other forms of aquaculture was discussed in Chapter 13. Salmon farming is now undertaken on a large scale in Canada and Chile, and there are reports of similar lice infestations as those experienced in northern Europe. The Labridae are one of the largest and most widespread families of fish (Chapter 1), and there may be other cleaner-species worthy of examination (Chapter 13). Cleaner-fish should not only be thought of in terms of salmon farming or ectoparasite control. Finfish mariculture is now widespread throughout the Mediterranean [e.g. sea-bass, *Dicentrarchus labrax* (L.), culture] and the possibility of using cleaner-fish is being examined. Even if the technique does not prove effective for other culture species, the use of cleaners to reduce net fouling may be possible. Use of cleaner-fish in shellfish culture should also not be dismissed.

The lack of fish of ideal size, disease introduction and transfer, over-exploitation of local populations, and the absence of suitable species in sufficient numbers, are problems associated with wrasse use which have been cited in this volume. It would appear that all of these concerns could be overcome by producing wrasse in hatchery breeding programmes. Pilot-scale projects have demonstrated that wrasse culture is feasible (Chapters 11 and 12) and perhaps these developments should be pursued if wrasse continue to be used as cleaner-fish.

The symposium on Wrasse Biology and Aquaculture Applications was called to bring together researchers from academia and the fish-farming industry, to summarize fundamental and applied wrasse research carried out over the past five to seven years and to discuss the future. The degree to which this was a successful exercise may be judged by the reader, but the evident success and continued use of a biological method of parasite control, and the associated stimulation in basic research into such a fascinating family of fishes, must be encouraging to all. The questions raised seem likely to stimulate further scientific investigation, and to provide the basis for the expansion of cleaner-fish technology into other areas of aquaculture.

Appendix 1
A guide to the use of cleaner-fish

(Modified from an article first published in *Fish Farmer* **17**(6): 7–8)

The following set of guidelines on the use of cleaner-fish in salmon farming were contributed to by the participants at the symposium *Wrasse Biology and Aquaculture Applications* held at Oban, Scotland, in October 1994.

1 Wrasse supply

(1) Only wrasse greater than 10 cm in length should be fished. Smaller wrasse are likely to escape from cages and this should ensure that young breeding adults are not fished.

(2) A specially designed collapsible wrasse trap is available from Norway. It is designed to minimize the capture of other animals which may attack wrasse, eat the bait, or interfere with sorting the catch.

(3) Ideally, wrasse would be fished from around the farm site, and not in areas where other salmon farms are situated.

(4) Samples of at least 30 wrasse per species should be screened for pathogens from the source population before stocking into salmon cages.

(5) Consider whether the vaccination of wrasse against *Vibrio* or furunculosis is necessary. This is recommended if it is anticipated that the wrasse will be exposed to these diseases on the farm.

(6) If possible, acclimatize wrasse to the net cage environment before stocking with salmon. Such acclimatization will bring forward the onset of cleaning of salmon when wrasse are stocked into cages.

(7) The industry would prefer wrasse to be cultured and certified as disease-free. Although initial trials have found the intensive culture of wrasse to be possible, a source of cultured wrasse is at present unavailable.

(8) Transport wrasse in dark, covered bins or tanks, with aeration if necessary.

(9) Never place ice in with wrasse as they may suffer salinity shock. Wrasse can survive warm temperatures (up to about 25°C) if sufficient oxygen is present.

(10) Do not supersaturate water with air or oxygen above 110% as this may stress the wrasse.

2 Stocking

(1) Aim to stock one wrasse to 50 salmon. Ratios of up to 1:150 have been found effective with Atlantic salmon smolts. Higher ratios may be necessary for larger salmon and during infestation by the louse *Caligus elongatus*.

(2) Cage nets should have a mesh size of 12–13 mm square. Less wrasse will be retained in larger mesh and very few in 18 mm mesh.

(3) Carefully release wrasse into cages below the surface by placing the transport bin underwater and allowing the wrasse to swim out against the cage net. Do not drop wrasse into a cage as this will provoke attack by the salmon.

(4) Stock wrasse as soon as possible after (or before) the input of smolts into sea-cages. It may take two weeks before lice levels are noticeably decreased by wrasse cleaning

3   Management

(1) Provide refuges for wrasse in cages. Bundles of used, narrow-diameter, plastic piping and old tyres have been found to be useful. Wrasse will sleep in these during darkness, and at low winter temperatures (less than 6°C); they may stay there and avoid chafing against the side of the net. Shelters may also provide a refuge from bird predation and attack from large salmon. Research into the optimal refuge design and placement is required.

(2) Wrasse presence and behaviour can be monitored by diver observation, underwater video, and recapture in traps. However, these methods will underestimate wrasse abundance and accuracy will vary with visibility, wrasse distribution, and salmon density.

(3) Wrasse need at least 4% of their body weight in food per day at summer temperatures (16°C). If less than this amount is available from lice and/or cage fouling, wrasse may be fed small or crushed mussels held in small mesh bags. Wrasse do not eat sufficient salmon feed to maintain good condition. Do not feed them shellfish from near (within a few 100 m) salmon farms which have a history of disease.

(4) Wrasse can be moved between cages in a cage-block to cope with variations in lice levels.

(5) Wrasse can be recaptured from a cage using traps, pots or creels. They may also be swum through from one cage to the next if salmon are transfered between cages by this method.

(6) Wrasse should not be moved between farm sites to minimise the potential of disease transfer.

(7) Remove wrasse from cages:

   (i)   before size-grading salmon;
   (ii) when starving salmon prior to harvest;
   (iii) if salinities of less than half-strength seawater (i.e. 15–17‰) are expected in the lower parts of cages at low temperatures (below 8°C).

(8) Corkwing and rock cook should be removed if temperatures below 6°C are expected. Goldsinny may be more tolerant of lower sea temperatures.

(9) Consider over-wintering wrasse in indoor tanks, or cages submersed at

depth. High (95–8%) winter survival of goldsinny has been achieved by maintaining them in wooden cages at 17 and 24 m depth.

(10)   Wrasse suffering from diseases should be destroyed in a humane and hygienic manner.

(11)   Wrasse remaining when all the salmon have been harvested should be quarantined and screened again before re-use. They should not be used again if either they, or the salmon they were stocked with, suffered any pathogenic infections.

4   Lice monitoring

(1)   Continue to monitor lice numbers on salmon after stocking with wrasse. Counting of all mobile (pre-adult and adult) and egg-bearing lice on each of at least 10 salmon per cage every week is recommended.

(2)   If lice levels increase:

    (i)   check cage nets for holes as wrasse may be escaping (this is the most common reason for failure of wrasse to control lice);

    (ii)   add more wrasse to cages;

    (iii) clean fouling off nets as wrasse may be feeding on it in preference to lice;

    (iv)   treat lice with a licensed chemotherapeutant (e.g. dichlorvos) under veterinary advice.

Wrasse seem to be less effective (but still useful) in controlling (a) *Caligus elongatus* compared to *Lepeophtheirus salmonis*; (b) lice on salmon in round compared with square cages; (c) lice on large (above 2 kg) salmon which have not been stocked with wrasse since they were smolts.

Index